Lecture Notes in Computer Science 15503

Founding Editors

Gerhard Goos
Juris Hartmanis

Editorial Board Members

Elisa Bertino, *Purdue University, West Lafayette, IN, USA*
Wen Gao, *Peking University, Beijing, China*
Bernhard Steffen, *TU Dortmund University, Dortmund, Germany*
Moti Yung, *Columbia University, New York, NY, USA*

The series Lecture Notes in Computer Science (LNCS), including its subseries Lecture Notes in Artificial Intelligence (LNAI) and Lecture Notes in Bioinformatics (LNBI), has established itself as a medium for the publication of new developments in computer science and information technology research, teaching, and education.

LNCS enjoys close cooperation with the computer science R & D community, the series counts many renowned academics among its volume editors and paper authors, and collaborates with prestigious societies. Its mission is to serve this international community by providing an invaluable service, mainly focused on the publication of conference and workshop proceedings and postproceedings. LNCS commenced publication in 1973.

Gwenolé Quellec · Mostafa El Habib Daho ·
Rachid Zeghlache
Editors

Image-Based Prediction of Retinal Disease Progression

MICCAI Challenges, DIAMOND 2024 and MARIO 2024
Held in Conjunction with MICCAI 2024
Marrakesh, Morocco, October 10, 2024
Proceedings

Editors
Gwenolé Quellec ⓘ
Inserm
Brest, France

Mostafa El Habib Daho ⓘ
Université de Bretagne Occidentale
Brest, France

Rachid Zeghlache ⓘ
Université de Bretagne Occidentale
Brest, France

ISSN 0302-9743　　　　　　ISSN 1611-3349 (electronic)
Lecture Notes in Computer Science
ISBN 978-3-031-86650-0　　ISBN 978-3-031-86651-7 (eBook)
https://doi.org/10.1007/978-3-031-86651-7

© The Editor(s) (if applicable) and The Author(s), under exclusive license to Springer Nature Switzerland AG 2025

This work is subject to copyright. All rights are solely and exclusively licensed by the Publisher, whether the whole or part of the material is concerned, specifically the rights of translation, reprinting, reuse of illustrations, recitation, broadcasting, reproduction on microfilms or in any other physical way, and transmission or information storage and retrieval, electronic adaptation, computer software, or by similar or dissimilar methodology now known or hereafter developed.
The use of general descriptive names, registered names, trademarks, service marks, etc. in this publication does not imply, even in the absence of a specific statement, that such names are exempt from the relevant protective laws and regulations and therefore free for general use.
The publisher, the authors and the editors are safe to assume that the advice and information in this book are believed to be true and accurate at the date of publication. Neither the publisher nor the authors or the editors give a warranty, expressed or implied, with respect to the material contained herein or for any errors or omissions that may have been made. The publisher remains neutral with regard to jurisdictional claims in published maps and institutional affiliations.

This Springer imprint is published by the registered company Springer Nature Switzerland AG
The registered company address is: Gewerbestrasse 11, 6330 Cham, Switzerland

If disposing of this product, please recycle the paper.

Preface

Diabetic retinopathy (DR) and age-related macular degeneration (AMD) are, together with cataracts and glaucoma, among the leading global causes of blindness in those aged 50 years and older. Over the years, to optimize their management, ophthalmologists have designed image-based algorithms (or scales) to predict their evolution in a patient. These algorithms rely on abnormalities visible in retinal images, either color fundus photography (CFP) or optical coherence tomography (OCT), that indicate increased risks of disease progression. They are based on simple population-level rules, which involve counting or measuring these abnormalities. As AI progresses, we expect that more fine-grained, patient-specific rules can be discovered for these predictive tasks, which may further improve the efficiency of DR or AMD patient management.

This led us to organize two challenges, which were accepted as satellite events of the 27th International Conference on Medical Image Computing and Computer Assisted Intervention (MICCAI 2024). These challenges aimed at deriving patient-specific rules for DR and AMD progression prediction from retinal images. The first challenge, Device-Independent diAbetic Macular edema ONset preDiction (DIAMOND), aimed at predicting the onset of center-involved DME, a vision-threatening complication of DR, one year in advance, for timelier treatments. The second challenge, Monitoring Age-Related macular degeneration progression In Optical coherence tomography (MARIO), aimed at predicting a significant progression of neovascular activity over the next three months, to optimize anti-VEGF treatments. These challenges relied primarily on data from French clinics; Given the known impact of domain shift on AI performance, we partnered with an Algerian clinic so that the domain generalizability of all algorithms could be evaluated.

The DIAMOND challenge relied on data from EviRed (*Evaluation Intelligente de la Rétinopathie Diabétique - Intelligent Evaluation of Diabetic Retinopathy*), a prospective cohort study involving up to 3,200 diabetic patients in France, utilizing multimodal imaging data—such as ultrawide-field CFP (UWF-CFP), OCT, and OCT angiography (OCTA)—to develop an artificial intelligence (AI)-powered prognostic tool for predicting the progression of DR. The goal was to assist ophthalmologists in decision-making, particularly in identifying patients at high risk for severe complications, such as proliferative DR and diabetic macular edema (DME), within a year of assessment. The DIAMOND challenge addressed one subtask of the EviRed project: it evaluated algorithms for predicting the onset of center-involved DME one year in advance, using UWF-CFP from various manufacturers. It was a fully blind challenge, where participants were required to submit both their training and evaluation code within a containerized environment.

The MARIO challenge targeted neovascular AMD (nAMD). Since 2007, anti-vascular endothelial growth factor (anti-VEGF) treatments have successfully slowed disease progression and improved visual function in nAMD patients. However, timely diagnosis, prompt treatment, and regular monitoring remain critical for optimal outcomes. The reliable detection of neovascular activity evolution using optical coherence tomography (OCT) is essential for individualized treatment strategies. The MARIO challenge evaluated algorithms for recognizing the evolution of neovascular activity in OCT scans, aiming to enhance anti-VEGF treatment planning. It included two tasks. The first task focused on pairs of 2D slices (B-scans) from two consecutive OCT acquisitions. The goal was to classify the evolution between these two slices (before and after), which clinicians typically examine side by side on their screens. The second task focused on the 2D slices level. The goal was to predict the future evolution within 3 months with close monitoring of patients that were enrolled in an anti-VEGF treatments plan. The dataset, collected from a single data center in Brest, France, provided metadata for each patient, including age, sex, number of visits, and intervals between visits. MARIO was an evaluation-only blind challenge, with the training dataset made available.

In both challenges, participants accessed the data and challenge resources via the Codabench platform (https://www.codabench.org/), agreeing to the challenge rules through a signed consent form. Submissions included detailed methodologies covering preprocessing, proposed approaches, pretraining, post-processing, results, and links to public code repositories. In MARIO, teams had the option to submit individual or combined papers for each task (only two teams opted for separate submissions). Selected teams also presented their work at a half-day satellite event during MICCAI 2024. This proceedings volume includes 15 papers from the MARIO challenge (17 submissions in total with 2 rejected) and 6 from the DIAMOND challenge (8 submissions in total with 2 rejected), showcasing a range of state-of-the-art deep learning approaches. Each paper underwent a single-blind peer review process involving at least two reviewers, focusing on novelty, clarity, formatting, and quality.

Organizing these two challenges together offered a unique opportunity to assess the impact of "blind training" on participation. A notable decline in the number of teams and submissions was observed in DIAMOND, which enforced blind training. While this trend was anticipated, it underscores the challenges in complying with regulations designed to protect patients, such as the European GDPR: as regulations around patient data become stricter, openly releasing healthcare data will become increasingly difficult in the future, which will necessarily impact future challenge design.

We extend our heartfelt gratitude to all participants, committee members, reviewers, and MICCAI organizers for their invaluable contributions. Their collective efforts have significantly advanced research in DR and AMD progression analysis and treatment planning.

January 2025

Gwenolé Quellec
Mostafa El Habib Daho
Rachid Zeghlache

Organization

General Chairs

Gwenolé Quellec   Inserm, France
Mostafa El Habib Daho   Université de Bretagne Occidentale, France
Rachid Zeghlache   Université de Bretagne Occidentale, France

Program Committee

Sarah Matta   Université de Bretagne Occidentale, France
Alireza Rezaei   Université de Bretagne Occidentale, France
Pierre-Henri Conze   IMT Atlantique, France
Matthias Gallardo   Fondation Rothschild, France
Mohammed Amine Lazouni   University of Tlemcen, Algeria
Mathieu Lamard   Université de Bretagne Occidentale, France
Ikram Brahim   Université de Bretagne Occidentale, France
Hicham Messaoudi   Université de Bretagne Occidentale, France

Additional Reviewers

Said Mahmoudi
Sidi Ahmed Mahmoudi
Amin Khaouani
Mohammed El Amine Bechar
Mohammed Ammar

Contents

DIAMOND Challenge

AI-Based Diagnostic Model for Predicting ci-DME Development 3
   Bo Yang, Yangyang Yan, and Wencheng Miao

Automatic Prediction of Center-Involved Diabetic Macular Edema Using
Ultra-Widefield Color Fundus Photography 20
   Philippe Zhang, Yihao Li, Weili Jiang, Jing Zhang, Sarah Matta,
   Yubo Tan, Hui Lin, Haoshen Wang, Jiangtian Pan, Hui Xu,
   Laurent Borderie, Alexandre Le Guilcher, Béatrice Cochener,
   Chubin Ou, Gwenolé Quellec, and Mathieu Lamard

A Device-Agnostic Deep Learning Approach for Predicting Ci-DME
Onset Using UWF-CFP Images 30
   Amin Khouani and Ihsane Mekki

Onset Prediction of Center-Involved Diabetic Macular Edema
on Ultra-Widefield Fundus Images Using Finetuned ResNet with Class
Balanced Focal Loss 41
   Yasine Lehmiani, Ayoub Elkhouzari, and Abdelhak Mahmoudi

Calibrated Models for DME Progression Prediction from Ultra-Wide Field
Retinal Images 54
   Adrian Galdran

Deep Learning Ensemble for Predicting Diabetic Macular Edema Onset
Using Ultra-Wide Field Color Fundus Image 64
   Pengyao Qin, Arun Thirunavukarasu, Theodoros Arvanitis, and Le Zhang

MARIO Challenge

Deep Learning Approaches for Monitoring Age-Related Macular
Degeneration Progression in Optical Coherence Tomography 77
   Yiding Hao

Leveraging MaxVit on Fused OCT Scan Pairs for Age-Related Macular
Degeneration Evolution Assessment 87
   Yosuke Yamagishi

Patch Progression Masked Autoencoder with Fusion CNN Network
for Classifying Evolution Between Two Pairs of 2D OCT Slices   97
   Philippe Zhang, Weili Jiang, Yihao Li, Jing Zhang, Sarah Matta,
   Yubo Tan, Hui Lin, Haoshen Wang, Jiangtian Pan, Hui Xu,
   Laurent Borderie, Alexandre Le Guilcher, Béatrice Cochener,
   Chubin Ou, Gwenolé Quellec, and Mathieu Lamard

Automatic Detection and Prediction of nAMD Activity Change in Retinal
OCT Using Siamese Networks and Wasserstein Distance for Ordinality 106
   Taha Emre, Teresa Araújo, Marzieh Oghbaie, Dmitrii Lachinov,
   Guilherme Aresta, and Hrvoje Bogunović

Exploring the Use of Off-the-Shelf AI Models for Complex Medical
Tasks: ResNet18 for Predicting Age-Related Macular Degeneration 118
   Amerens A. Bekkers, Nina M. van Liebergen, and Hugo J. Kuijf

AI-Driven Analysis of Sequential OCT Images for Detecting Neovascular
Activity in Age-Related Macular Degeneration to Optimize Anti-VEGF
Therapy 126
   Yi Ding

Efficient Deep Learning Models for Evaluating the Progression
of Age-Related Macular Degeneration Through Optical Coherence
Tomography 135
   Abdul Qayyum, Moona Mazher, Imran Razzak, and Steven A. Niederer

Optimizing Anti-VEGF Treatment Strategies with AI-Based Neovascular
AMD Detection 145
   Yi Ding

Monitoring Age-Related Macular Degeneration Progression in Optical
Coherence Tomography (MARIO), Task 1 - MICCAI Challenge 2024,
jkulinzstudents Submission 154
   Marcel Huber, Patrick Binder, and Markus Frohmann

Monitoring Age-Related Macular Degeneration Progression in Optical
Coherence Tomography (MARIO), Task 2 - MICCAI Challenge 2024,
jkulinzstudents Submission 163
   Patrick Binder, Marcel Huber, and Markus Frohmann

Modular Transformer-Based Monitoring and Prediction of Wet AMD
Progression Using OCT Imaging 172
   Lovre Antonio Budimir, Ivana Matovinović, Donik Vršnak,
   and Sven Lončarić

A Novel Multimodal Deep Learning Fusion Framework for Predicting
Neovascular Activity Evolution in Exudative Age-Related Macular
Degeneration 182
   *Christopher Nielsen, Ahmad O. Ahsan, Matthias Wilms,
   and Nils D. Forkert*

Evaluating Pretraining Strategies for OCT-Based Macular Degeneration
Classification 192
   *Jessica Kächele, Robin Peretzke, Alexandra Ertl, Maximilian Fischer,
   Marlin Hanstein, Florian Max Hauptmann, Dimitrios Bounias,
   Marco Nolden, Peter Neher, and Klaus H. Maier-Hein*

Classification and Prediction of Age-Related Macular Degeneration
Progression Using OCT Images and Multiple Instance Learning 201
   *Alberto J. Beltrán-Carrero, Javier Torresano-Rodríguez,
   Esther Santos-Vicente, María J. Aparicio Hernández-Lastras,
   Álvaro Caballero-Sastre, María J. Ledesma-Carbayo,
   and Juan J. Gómez-Valverde*

Multi-modal Siamese Ensemble for Neovascular AMD Classification
and Prediction from Optical Coherence Tomography 211
   Sebastien Richard and Marie Beurton-Aimar

Author Index 223

DIAMOND Challenge

AI-Based Diagnostic Model for Predicting ci-DME Development

Bo Yang[1,2(✉)] and Yangyang Yan[2]

[1] Wonderful Things Lab, No. 3 Xingfu Street, Beifang Town, Huairou Science City, Huairou District, Beijing 101407, China
yeungbo@gmail.com
[2] AIFUTURE Lab, Shenzhou Digital Building, No. 16 Suzhou Street, Haidian District, Beijing 100085, China

Abstract. This research is centered on the development of an AI-based diagnostic model for central-involved diabetic macular edema (ci-DME) using ultra-widefield color fundus photography (UWF-CFP). The model's training and validation datasets were constructed from data collected under the EviRed project framework at 14 hospitals in France, supplemented with independent data from the LAZOUNI Ophthalmic Clinic in Tlemcen, Algeria. The development process entailed the construction of a comprehensive and robust dataset, coupled with the implementation of advanced optimization techniques. The results of this study indicate that both EfficientNet-B7 and ResNet152, with specially designed training strategies are effective models for the diagnosis of ci-DME. EfficientNet-B7 demonstrated superior overall discrimination ability, while ResNet152 showcased strong classification performance and reliable confidence predictions. These findings highlight the significant potential of AI in advancing the diagnosis and treatment of ci-DME, potentially facilitating earlier interventions and resulting in improved patient outcomes.

Keywords: Central-involved diabetic macular edema (ci-DME) · Ultra-widefield color fundus photography (UWF-CFP) · Artificial Intelligence (AI)

1 Introduction

Diabetic macular edema (DME) is a significant and leading cause of vision loss among individuals with diabetes globally. Characterized by the accumulation of fluid in the central portion of the retina, DME can occur at any stage of diabetic retinopathy (DR) and poses a substantial threat to patients' visual acuity. Early detection and management of DME, particularly the center-involved form (ci-DME), are crucial for preventing irreversible vision impairment. According to research by Zhang et al. [1], ci-DME is the primary cause of significant vision loss. Currently, the presence of ci-DME is primarily assessed using three-dimensional optical coherence tomography (OCT) imaging. Recent studies have indicated that two-dimensional color fundus photography (CFP) can also be used to evaluate ci-DME. [2] With the advent of ultra-widefield (UWF) imaging technology, it has overcome the field-of-view limitations of traditional CFP, achieving a single-capture angle of up to 200° and covering approximately 80% of the total retinal area. Compared to conventional color fundus photography (CFP), ultra-widefield (UWF) imaging provides a broader field-of-view, non-mydriatic capability, and high resolution. According to Vujosevic et al. [3], observing peripheral retinal lesions (PPL) within UWF images is instrumental in the early diagnosis of center-involved diabetic macular edema (ci-DME) and facilitates timely intervention, demonstrating the promising application prospects of UWF imaging.

This study aims to revolutionize the diagnostic and treatment approach of ci-DME by applying artificial intelligence (AI) and deep learning technologies to the latest advancements in ultra-widefield color fundus photography (UWF-CFP). The challenge lies in developing and evaluating a model that can predict whether a patient will develop ci-DME within one year, based solely on UWF-CFP images. Given that UWF-CFP provides a broader field-of-view, we hypothesize that it is more likely to capture early signs of DME, as shown in Fig. 1. If successful, this study will significantly enhance early detection and treatment planning, reduce the incidence of blindness, and demonstrate the value of AI in healthcare.

In the realm of artificial intelligence (AI) and deep learning technologies applied to research on center-involved diabetic macular edema (ci-DME), the Codabench platform serves as a valuable asset. Codabench, recognized for its flexibility, ease of use, and reproducibility, offers a meta-benchmark platform [4] that can be harnessed for a spectrum of machine learning tasks, with diabetic retinopathy detection being a key application. The general deep learning model for detecting diabetic retinopathy, as proposed by Chen et al. [5], and the autonomous AI-based diagnostic system for primary care offices by Abràmoff et al. [6], collectively highlight AI's pivotal role in predicting diabetic retinopathy (DR). Furthermore, the comprehensive review on deep learning in ophthalmology by Ting et al. [7] delves into the technical and clinical nuances of the field, including the critical aspect of DR detection.

To promote the technical level of UWF-CFP in ci-DME analysis, we participated in the DIAMOND Challenge. This challenge focuses on developing and evaluating a model that can predict whether a patient will develop ci-DME within a year, based solely on UWF-CFP images. Additionally, the DIAMOND Challenge introduces a new approach different from typical predictive modeling competitions. Participants will not have access to the training, validation, or test datasets, and hence, will not have the opportunity to train their models directly. Instead, participants need to submit their code, which will be encapsulated in Docker containers by the organizers. The organizing committee will run these codes on a specialized cloud-based cluster. This unique approach simulates a

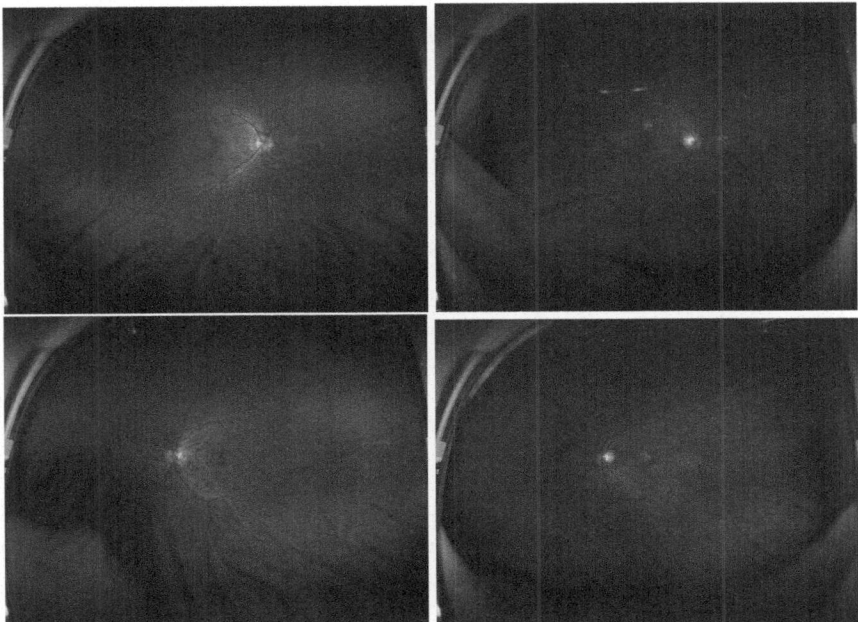

Fig. 1. Examples of UWF fundus images depicting 'Non-ci-DME' on the left and 'ci-DME' on the right.

real-world scenario where data accessibility is often restricted due to privacy regulations, ethical considerations, or logistical issues.

Once participants submit their code, the challenge committee will run it in the organizers' environment, using their proprietary dataset for model training. The training data comprises data collected from 14 hospitals in France under the EviRed project framework (https://evired.org/). EviRed not only focuses on predicting the progression of ci-DME but also has a broader goal of predicting general diabetic retinopathy (DR) complications. Preliminary experiments using a simple baseline algorithm (ResNet-50) on EviRed data have shown that predicting ci-DME occurrence using UWF-CFP is feasible (area under the ROC curve = 0.73). For performance evaluation, the challenge committee will also use independent data from the LAZOUNI Ophthalmic Clinic in Tlemcen, Algeria. By leveraging multiple datasets, including data from French and Algerian populations and images captured using different UWF-CFP devices, the challenge aims to develop a solution that transcends geographical and demographic boundaries. This universality is vital for ensuring that the predictive models developed are robust and effective across different populations, enhancing their clinical utility globally.

2 Dataset

2.1 Data Source

1. The European dataset includes images acquired using the CLARUS (CLARUS 500, Carl Zeiss Meditec Inc.) and OPTOS (P200Dx, Optos plc) devices.
2. The African dataset includes images acquired using the OPTOS (P200Dx, Optos plc) and EIDON (EIDON, iCare, Revenio Group) devices.

These images were obtained by experienced orthoptists at specialized ophthalmic clinics. The European data used for training, validation, and testing were collected from 14 partner hospitals in France, all of which are participants in the EviRed project. This dataset encompasses various data points essential for our analyses. As for the African data used for testing, these images were collected from the LAZOUNI Ophthalmic Clinic located in Tlemcen, Algeria (https://clinique-lazouni.business.site/).

Each case refers to a single UWF-CFP image. All training, validation, and testing cases are labeled based on the presence or absence of ci-DME (central-involved diabetic macular edema) within one year of capturing the UWF-CFP image.

2.2 Blind Dataset with Real Dataset

The dataset of the competition was derived from the EviRed cohort, comprising 7,065 cases from 1,917 patients across 2,672 annual visits (with each patient potentially having one to three visits). During each visit, the patients' eyes were typically imaged using one or two imaging devices (CLARUS and/or OPTOS). Additionally, the study incorporated 300 cases from private clinics in Algeria, involving 150 patients (with each patient having only one visit and a single image per eye). At the time of the visit, each eye was imaged using either the OPTOS or EIDON device.

In terms of dataset partitioning, the training set included 5,652 cases from the EviRed cohort, the validation set contained 707 cases, and the test set comprised 706 cases from the EviRed cohort as well as all 300 cases from private clinics in Algeria. Within the EviRed cohort, all images from the same patient were allocated to the same subset (training, validation, or testing). The proportion of positive cases in the EviRed cohort was 7.5%, while the proportion of positive cases in the Algerian dataset will be disclosed after the challenge to ensure unbiased blind assessment of an unseen population.

The dataset was divided following the standard 80:10:10 ratio to ensure balance during model development. The training set included images from two different devices (OPTOS and CLARUS) within the same cohort to increase data variability. The test set used images from the same cohort to maintain consistency. Furthermore, the study collected images from the OPTOS device across different countries and continents and introduced a new device, EIDON, which was not involved in training. This strategy aims to encourage participants to develop device-agnostic models that can accurately predict the occurrence of center-involved diabetic macular edema (ci-DME) independent of population and imaging equipment.

The annotations of the aforementioned dataset were meticulously completed by ophthalmologists with at least 10 years of professional experience, based on clinical visit information collected over a year. This clinical information included OCT (Optical Coherence Tomography) images and visual acuity data of the same eye. Throughout the annotation process, ophthalmologists focused on considering center-involved diabetic macular edema (DME). To ensure consistency and reliability of the annotations, the annotation process for the training, validation, and testing sets all adhered to the same standards. The annotation work in the African centers also strictly followed this protocol. To de-identify and ensure correct RGB encoding, all images have been converted from DICOM to PNG format.

2.3 Local Training Data with Synthetic Dataset

To locally develop and verify algorithmic models, participants are required to prepare their own datasets for running algorithms. Thus, we utilized the dataset from the Deep-Diabetic-Retinopathy-Image-Dataset (DeepDRiD) project [8] for local training and debugging. This project provides 256 ultra-widefield fundus images along with the original labels for the 5-class diabetic retinopathy classification. Based on the conversion relationship between the "5-class diabetic retinopathy" labels and the "binary classification of development into macular edema within one year" labels provided by the competition organizing committee, we were able to obtain the corresponding "binary classification of development into macular edema within one year" labels for the 256 images provided by DeepDRiD. Additionally, Task 2 of the Ultra-Widefield Fundus Imaging for Diabetic Retinopathy Challenge 2024 (UWF4DR) offered 201 ultra-widefield fundus images, accompanied by "referable diabetic retinopathy binary classification" labels.

Referring to the definition of referable diabetic retinopathy, which is defined as moderate non-proliferative diabetic retinopathy (NPDR) or worse, including diabetic macular edema (DME), and using the conversion relationship between the "5-class diabetic retinopathy" labels, "binary classification of development into macular edema within one year" labels, the "referable diabetic retinopathy" labels were transformed into "development into macular edema within one year" labels, while the "non-referable diabetic retinopathy" labels were converted into "no development into macular edema within one year" labels. Utilizing this approach, we compiled a local synthetic training dataset for our model by incorporating data from both DeepDRiD and UWF4DR, resulting in a total of 407 images, as detailed in Table 1. Of these, there are 221 images labeled as "development into macular edema within one year" and 236 images labeled as "no development into macular edema within one year." The dataset was divided into training and validation sets in an 8:2 ratio.

Table 1. Synthetic Dataset Derived from DeepDRiD and UWF4DR with Converted Labels

Data Source	ci-DME	Non ci-DME	Total number
DeepDRiD	109	147	256
UWF4DR	112	89	201
Total number	221	236	407

Moreover, to enrich the training dataset and enhance the performance and generalization capabilities of the model, we have implemented a series of data augmentation operations aimed at improving the model's robustness and adaptability: ultra-widefield fundus images were read using Pillow.

2.4 Data Augmentation

The data augmentation techniques applied to the training set are crucial for enhancing the model's generalization capabilities and improving its robustness against variations in input data:

Size Scaling: By resizing the cropped images to a uniform dimension of 448 × 448 pixels, the model is exposed to a consistent input size. This consistency is vital for the convolutional layers to effectively learn features without being biased by the size of the input data. It also ensures that the network can handle inputs regardless of their original dimensions, as long as they are scaled accordingly.

Random Flipping: Introducing randomness by flipping images horizontally and vertically simulates the variability in orientation that may be encountered in practical scenarios. This augmentation helps the model to become invariant to orientation, thus improving its ability to recognize patterns regardless of their spatial orientation in the image.

Random Rotation: Limiting the rotation angle to within 15° provides a slight augmentation that accounts for minor image misalignments or rotations without introducing extreme variations that could confuse the model. This subtle augmentation aids in the model's ability to recognize features under small rotational changes, contributing to its overall generalization.

Image Normalization: Normalizing the image data to a mean of [0.485, 0.456, 0.406] and a standard deviation of [0.229, 0.224, 0.225] aligns the input data distribution with the expected distribution of the model's layers. This step is essential for stabilizing the learning process, as it reduces the impact of outliers and speeds up convergence by ensuring that the input data is on a similar scale.

For the validation set, a slightly different process is employed, focusing on:

Image Size Scaling: Maintaining the consistency in image dimensions for the validation set is crucial for making a fair assessment of the model's performance. The 448 × 448 pixels size ensures that the validation data mimics the training data in terms of input size.

Image Normalization: Using the same mean and standard deviation as the training set ensures that the validation data is treated in the same way, allowing for an accurate evaluation of the model's performance without any bias introduced by different normalization parameters.

The impact of these data augmentation techniques on model performance is significant. They contribute to a more robust and accurate model by exposing it to a variety of transformations that it may encounter in real-world applications. This exposure to diverse data during training leads to better generalization, potentially increasing the model's accuracy and reducing overfitting. The careful application of these techniques ensures that the model learns to handle inputs in a way that closely mirrors the complexity and variability of real-world data.

2.5 Evaluation Metrics

In this study, evaluation metrics consistent with the predictive targets of high sensitivity and specificity were employed, including the Area Under the Curve (AUC), which plots the true positive rate against the false positive rate at various threshold settings. This metric is commonly used to assess medical classification/prediction challenges and competitions on platforms such as Grand Challenge and Kaggle, as well as in medical literature. Two secondary metrics (with lower weights) were considered to further evaluate the correlation of computed probabilities:

The use of F1 ensured the relevance of binary predictions using a probability cutoff value of 0.5, which is the harmonic mean of precision and recall, providing a single measure that balances both.

The use of Expected Calibration Error (ECE) ensured the calibration properties of the model, which measures the difference between the predicted probabilities and the observed frequencies of outcomes. It assesses how well the model's predicted probabilities align with the true distribution of labels.

The final ranking of the submitted algorithms will be determined by a composite score reflecting their performance across four distinct test subsets (D1: CLARUS/European population, D2: OPTOS/European population, D3: OPTOS/African population, and D4: EIDON/African population). This approach aims to evaluate the device-independence and population-independence capabilities of the proposed methods. Each algorithm will be evaluated separately on the four test subsets. The performance of each subset D will be measured using the selected metric S(D) (S = AUC + 0.5 × F1 + 0.5 × (1 − ECE)). To derive the final score for each submission, the organizing committee will average the individual performance scores: Final Score = S(D1) + S(D2) + S(D3) + S(D4). It is noteworthy that, although each test subset is equally weighted, a greater emphasis is placed on out-of-domain test cases, as the corresponding test subsets (D3 and D4) are smaller.

Testing is conducted by the organizing committee on a closed dataset and within a test environment. Each algorithm will provide a single prediction for each case. If a prediction is missing for a test case, a default probability of 0 will be assigned (indicating "no ci-DME within one year after UWF-CFP image capture"). The rationale behind this method is to ensure that algorithms demonstrating consistent performance across various devices and populations receive higher rankings, thereby highlighting their robustness and generalizability.

To facilitate a more robust statistical analysis of algorithm performance, the Delong test was also employed to statistically compare the ROC curves of competing models, providing a way to determine if there are significant performance differences between them. It enables us to statistically determine whether there are significant performance differences between these approaches.

In summary, these evaluation metrics are chosen for their ability to provide a comprehensive assessment of the predictive models' performance in the context of ci-DME diagnosis. They not only measure the accuracy of the models but also their reliability and generalizability across different populations and devices, which is crucial for the practical application of these models in a clinical setting.

3 Methodology

3.1 Method Pipeline

As depicted in Fig. 2, we have developed an AI-based diagnostic model for central-involved diabetic macular edema (ci-DME) utilizing ultra-widefield color fundus photography (UWF-CFP) through a six-step methodology. Initially, we devised a method for label conversion to prepare a synthetic dataset derived from DeepDRiD and UWF4DR. Subsequently, we applied data augmentation techniques to expand the dataset, exposing the synthetic dataset to a variety of transformations it is likely to encounter in real-world applications. Thirdly, models were trained on this augmented local synthetic dataset to produce a pre-trained model. In the fourth step, the training code and the pre-trained model were submitted to dedicated servers provided by OVHcloud for further training with a restricted training dataset. The fifth step involved evaluating the trained model using a closed test dataset, followed by ranking on a leaderboard.

3.2 Computing Environments

To ensure fairness in resource allocation, each participant was provided with an identical computing environment, whether for training or performance evaluation. The algorithms were executed on dedicated servers provided by OVHcloud (https://www.ovhcloud.com) and funded by the EviRed project. The training, validation, and testing datasets were also hosted on OVHcloud, with computations taking place on machines equipped with NVIDIA Tesla V100S GPUs. This setup guaranteed that all participants had access to equal computational resources.

3.3 Selection of AI Model Architectures

ResNet152. Initially, we selected the ResNet152 (Residual Network) [9] as the base network architecture, which is a deep convolutional neural network (CNN) proposed by Kaiming He et al. from Microsoft Research in 2015. It has

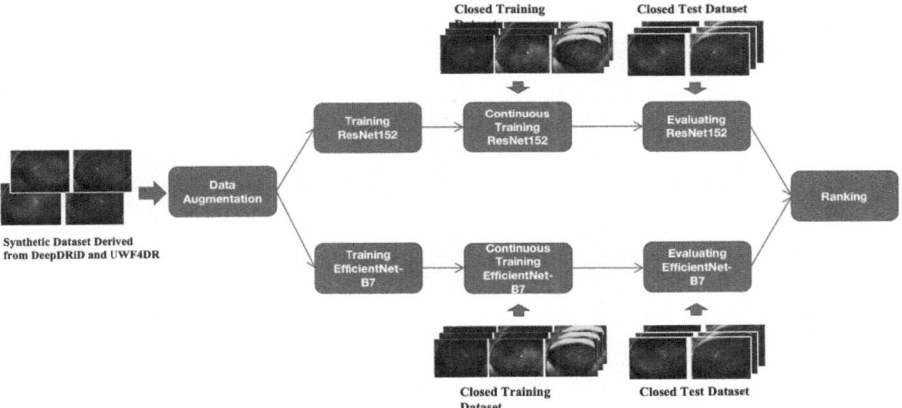

Fig. 2. The Pipeline of AI-based Diagnostic Model for Predicting ci-DME.

demonstrated exceptional performance in image recognition and classification tasks. Figure 3 illustrates the internal structure of a residual block within the ResNet-152 architecture. The block begins with a 1×1 convolutional layer with 64 filters, followed by a ReLU activation function. Subsequently, a 3×3 convolutional layer with 64 filters and stride 1 is applied, again followed by a ReLU activation. A third 1×1 convolutional layer with 256 filters is then used to expand the dimensionality, leading to a summation with the input (skip connection) to preserve spatial information. Finally, the output is passed through a ReLU activation function. The '256-d' denotes the dimensionality of the input data, where 'd' is a variable representing the depth of the input feature map. This design is pivotal for ResNet-152, enabling the model to learn deep representations efficiently while mitigating the risk of overfitting.

EfficientNet-B7. The EfficientNet-B7 architecture [10], lauded for its efficiency and generalization, underpins our model. To boost training efficiency and performance, we introduced a Dropout layer with a 0.5 rate and a linear output layer, initializing with pre-trained ImageNet-1K weights. This setup allows swift convergence with minimal data, leveraging pre-trained visual feature knowledge. Figure 4 illustrates the detailed structure of a MBConv (Modulated Batch Normalization Convolution) block, a fundamental component of the EfficientNet-B7 model. The block begins with a 1×1 convolutional layer for dimensionality reduction, followed by batch normalization and the Swish activation function. Subsequently, a depthwise separable convolution with a kernel size of 3×3 or 5×5 is applied, which can be adjusted depending on stride (s1 or s2). The output is then subjected to another round of batch normalization and Swish activation. A Squeeze-and-Excitation (SE) module is integrated, featuring global average pooling, two fully connected layers (FC1 and FC2), and a sigmoid activation function to recalibrate channel-wise feature responses. The processed features are then

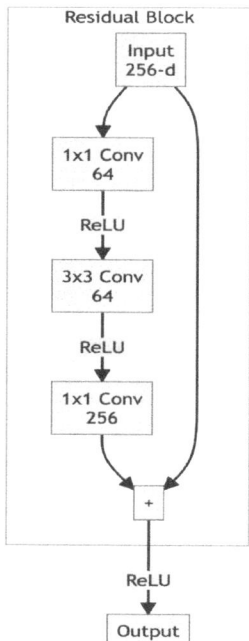

Fig. 3. Schematic of a Residual Block in the ResNet-152 Architecture

convolved through a 1 × 1 convolution, batch normalized, and optionally passed through a dropout layer before being added back to the input via a skip connection, facilitating the preservation of spatial information and enhancing the model's ability to learn complex representations.

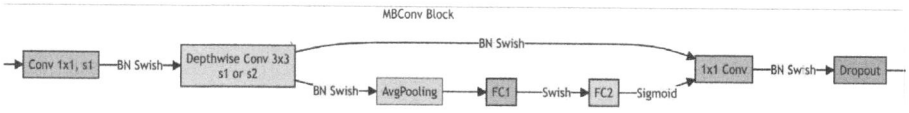

Fig. 4. Schematic Representation of the MBConv Block within the EfficientNet-B7 Architecture

4   Training Strategies and Evaluation

4.1   Strategy I

The model was trained on a local training dataset using the Adam optimizer with a learning rate of 0.0001, a batch size of 32, and a maximum epoch of

100. The training process monitored the validation set score S, defined as S = AUC + 0.5 × F1 + 0.5 × (1 − ECE). The best-performing model on the validation set during the training process was saved as the initial weight for training in the closed environment of the organizer's restricted dataset. During training, the Adam optimizer was still employed with a learning rate of 0.0001, a batch size of 16, and a maximum epoch of 100. The training process continued to monitor the validation set score S, calculated as S = AUC + 0.5 × F1 + 0.5 × (1 − ECE), and the best-performing model on the validation set was saved. Additionally, the model's prediction results on the validation set were output. The training monitoring metrics, including loss, AUC (Area Under the Curve), F1 (F1 Score), ECE (Expected Calibration Error), and S (composite score), as derived from the organizer's feedback logs, are depicted in Fig. 5

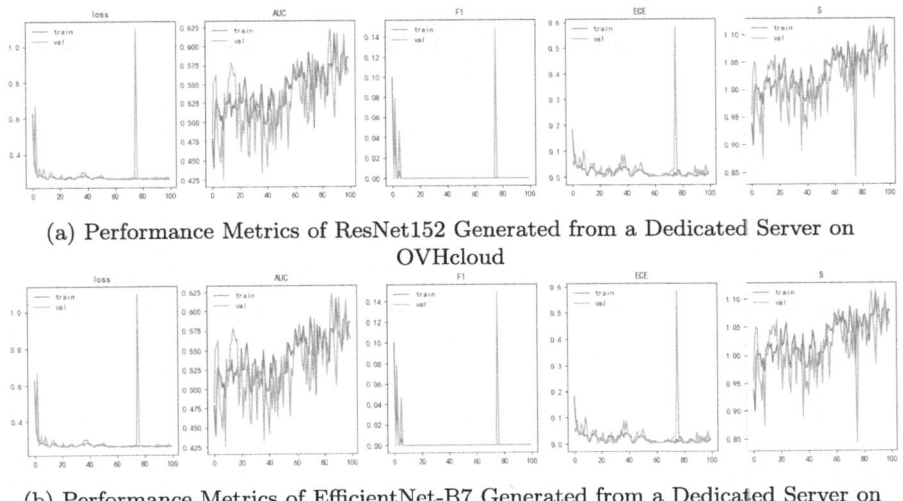

(a) Performance Metrics of ResNet152 Generated from a Dedicated Server on OVHcloud

(b) Performance Metrics of EfficientNet-B7 Generated from a Dedicated Server on OVHcloud

Fig. 5. Training Monitoring Metrics for Strategy I

4.2 Strategy II

During training on a local training dataset, the Adam optimizer was utilized with a learning rate set to 0.0001, a batch size of 32, and a maximum epoch of 100. The training process monitored the validation set score S, defined as S = AUC + 0.5 × F1 + 0.5 × (1 − ECE). The model with the best performance on the validation set during the training process was saved as the initial weight for training within the organizer's closed and restricted dataset environment. When training on the closed dataset, the loss function calculation referred to error values from both the training and validation sets. The training employed the Adam optimizer with a learning rate of 0.0001, a batch size of 16, and a maximum epoch of 100. The training process continued to monitor the validation set score S, calculated as S = AUC + 0.5 × F1 + 0.5 × (1 − ECE), and the best-performing model on the validation set was saved. The model's prediction results on the validation set were also output. The training monitoring metrics, including loss, AUC, F1, ECE, and S, as derived from the organizer's feedback logs, are presented in Fig. 6.

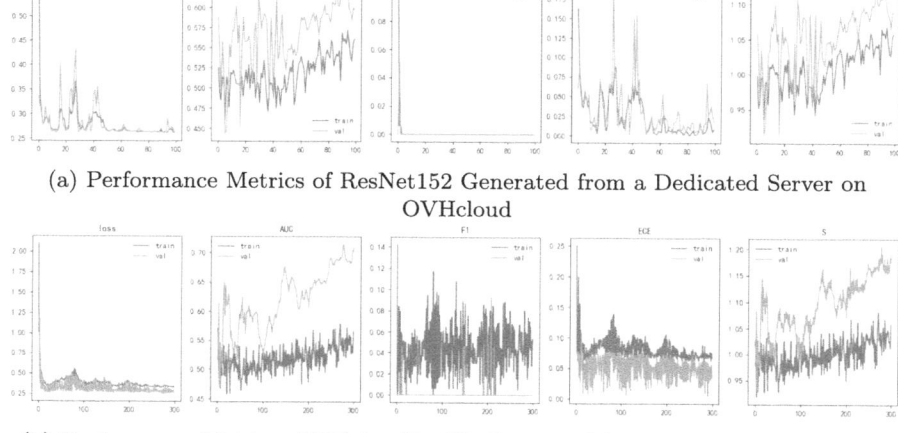

(a) Performance Metrics of ResNet152 Generated from a Dedicated Server on OVHcloud

(b) Performance Metrics of EfficientNet-B7 Generated from a Dedicated Server on OVHcloud

Fig. 6. Training Monitoring Metrics for Strategy II

4.3 Strategy III

In the training process conducted on local datasets, the loss function was corrected by referencing error values from both the training and validation sets. The Adam optimizer was utilized with an initial learning rate of 0.0001, and a cosine annealing learning rate schedule was implemented. The batch size was set to 32, and the maximum number of epochs was set to 100. The training process monitored the validation set score S, defined as S = AUC + 0.5 × F1 + 0.5 × (1 − ECE). The model with the highest performance on the validation set during the training phase was saved as the initial weight for subsequent training within the organizer's closed and restricted dataset environment.

During training on the closed dataset, the real-world training dataset was restructured by combining a segment of the training set with the entire validation set. This reconfiguration resulted in a data distribution where the volume of data labeled as "will not develop macular edema within one year" was twice that of the data labeled as "will develop macular edema within one year," while the validation set remained unchanged. The training process continued to employ the Adam optimizer with an initial learning rate of 0.0001, a ccsine annealing learning rate schedule, a batch size of 16, and a maximum number of epochs of 100. The training process monitored the validation set score S, calculated as S = AUC + 0.5 × F1 + 0.5 × (1 − ECE). The best-performing model on the validation set was saved, and the model's prediction results on the validation set were output. The training monitoring metrics, including loss, AUC, F1, ECE, and S, as reported in the organizer's feedback logs, are illustrated in Fig. 7.

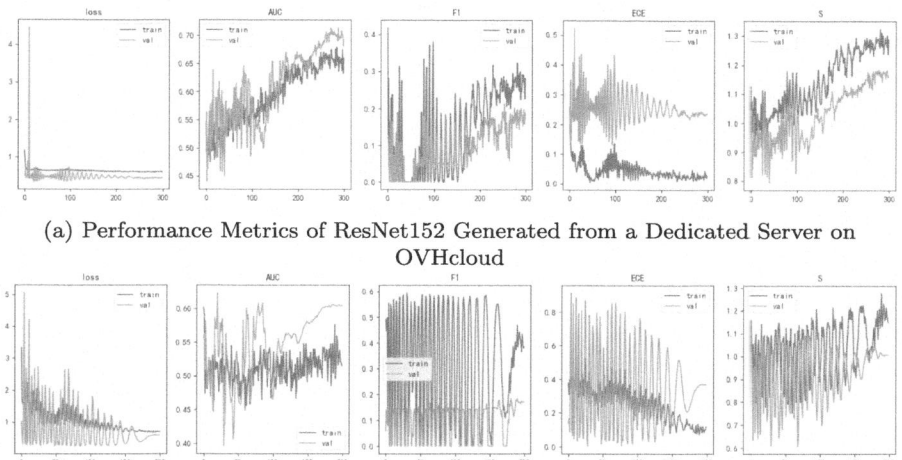

(a) Performance Metrics of ResNet152 Generated from a Dedicated Server on OVHcloud

(b) Performance Metrics of EfficientNet-B7 Generated from a Dedicated Server on OVHcloud

Fig. 7. Training Monitoring Metrics for Strategy III

5 Discussion

Table 2. Performance Evaluation Outcomes for the Deep Learning Models EfficientNet-B7 and ResNet152 Across Three Distinct Strategic Apprcaches

Heading level	Model	Epoch	Score	AUC	F1 Score	ECE
Strategy I	EfficientNet-B7	100	1.121	0.6625	0.1486	0.0829
	ResNet152	100	1.1152	0.6233	0.0000	0.0162
Strategy II	EfficientNet-B7	300	**1.2049**	**0.7145**	0.0000	0.0191
	ResNet152	100	1.1304	0.6329	0.0000	**0.0049**
Strategy III	EfficientNet-B7	200	1.0912	0.6087	0.0000	0.0349
	ResNet152	300	**1.1836**	0.7065	**0.1891**	0.2351

Table 2 presents the performance evaluation results of two deep learning models, EfficientNet-B7 and ResNet152, under three distinct strategies. The following analysis compares the models' performance and discusses the study's limitations.

5.1 Comparative Model Performance

EfficientNet-B7. Demonstrates superior performance in Strategies I and II, particularly in AUC and overall score. Shows relatively weaker performance in Strategy III. Consistently yields an F1 score of 0 across all strategies, potentially indicating a class imbalance issue.

ResNet152. Exhibits optimal performance in Strategy III, achieving the highest F1 score (0.1891).

Maintains lower ECE values across all strategies, suggesting more reliable confidence predictions. Slightly underperforms compared to EfficientNet-B7 in Strategies I and II.

5.2 Inter-Strategy Comparison

Strategy I: EfficientNet-B7 outperforms ResNet152 in AUC and F1 score.
ResNet152 demonstrates superior performance in ECE.

Strategy II: EfficientNet-B7 achieves the highest overall score (1.2049) and AUC (0.7145).
ResNet152 maintains a lower ECE value (0.0049 vs. 0.0191).

Strategy III: ResNet152 excels in AUC (0.7065) and F1 score (0.1891).
EfficientNet-B7 exhibits a lower ECE value (0.0349 vs. 0.2351).

5.3 Key Findings

The choice of strategy significantly impacts model performance. EfficientNet-B7 demonstrates superior overall discriminative ability. ResNet152 shows stronger classification capability and more reliable confidence predictions. Consistently low F1 scores across all scenarios suggest a severe class imbalance issue.

5.4 Study Limitations

Dependence on Synthetic Data: The study's reliance on synthetic data may lead to discrepancies between model performance in simulated environments and real-world scenarios. While synthetic data can provide large sample sizes, it may fail to fully capture the complexity and variability of real-world data.

Limited Access to Actual Challenge Dataset: Insufficient access to the actual challenge dataset may constrain the models' generalization capabilities, potentially resulting in suboptimal performance when faced with novel, unseen data.

Class Imbalance Issues: The consistently low F1 scores, particularly EfficientNet-B7's score of 0 across all strategies, strongly indicate a class imbalance problem. This imbalance may lead to biased predictions for certain classes, affecting overall model performance.

Impact of Strategy Selection: Although the research demonstrates that different strategies significantly affect model performance, it lacks an in-depth exploration of the specific implementation details of these strategies and their underlying mechanisms influencing model behavior. From Table 2, it is evident that the choice of different strategies significantly impacts model performance. EfficientNet-B7 excels in overall discrimination ability, whereas ResNet152 shows stronger classification capability and more reliable confidence predictions. Finally, we achieved second place and third place with our strategies on the leaderboard for ci-DME competitions, as shown in Table 3 below.

Table 3. Leaderboard Results of Ranking for DIAMOND Challenge.

Participant	Score	AUC	F1 Score	ECE
agaldran	1.2184	0.7203	0.0454	0.0492
yangbo	1.2049	0.7145	0.0000	0.0191
yangbo	1.1836	0.7065	0.1891	0.2351

6 Conclusion

The study meticulously contributes to the field of ophthalmology by introducing an AI-driven approach for diagnosing center-involved diabetic macular edema (ci-DME) using ultra-widefield color fundus photography (UWF-CFP). Our findings robustly indicate that both EfficientNet-B7 and ResNet152 architectures serve as proficient models in the diagnostic process of ci-DME. Notably, EfficientNet-B7 distinguishes itself with superior discrimination capabilities, particularly excelling in Area Under the Curve (AUC) metrics under Strategies I and II, and peaking with the highest score under Strategy II. This underscores its efficacy in discerning subtle patterns indicative of ci-DME.

Conversely, ResNet152 demonstrates a commendable proficiency in Expected Calibration Error (ECE), achieving the pinnacle in both AUC and F1 scores under Strategy III. This suggests that ResNet152 not only classifies with high precision but also provides dependable confidence predictions, which is pivotal for clinical decision-making.

The implications of our study are profound, highlighting the transformative potential of AI in refining the diagnostic process and treatment strategies for ci-DME. The prospective for earlier interventions and augmented patient outcomes is significant, advocating for a paradigm shift towards proactive management of the condition.

Our research advances by proposing future directions, including the integration of additional imaging modalities and multi-modal data. This approach has the potential to further enhance the model's accuracy, aligning with the burgeoning trend in medical imaging towards more holistic diagnostic methods. Furthermore, we advocate for extending our study to evaluate the model's generalizability across various populations and clinical settings. Such an extension could yield diagnostic tools that are not only more inclusive but also adaptable to a broader range of clinical scenarios.

In comparison with existing literature, our study stands out by its comprehensive approach and the novel application of advanced AI models to ci-DME diagnostics. Unlike other works that may focus on single strategies or models, we provide a comparative analysis, demonstrating the strengths of each model within the context of ci-DME diagnosis. This comparative evaluation is crucial for the field, as it guides future research and clinical applications towards the most effective and reliable diagnostic methods.

Acknowledgments. We would like to extend our sincere gratitude to the experts who provided invaluable medical expertise and technical support throughout our research work. Their knowledge and guidance played a crucial role in shaping the direction and improving the quality of our study.

References

1. Zhang, J., et al.: Diabetic macular edema: current understanding, molecular mechanisms and therapeutic implications. Cells **11**(21), 3362 (2022)
2. Varadarajan, A.V., Bavishi, P., Ruamviboonsuk, P., et al.: Predicting optical coherence tomography-derived diabetic macular edema grades from fundus photographs using deep learning. Nat. Commun. **11**(1), 130 (2020). https://doi.org/10.1038/s41467-019-13922-8
3. Vujosevic, S., et al.: Screening for diabetic retinopathy: new perspectives and challenges. Lancet Diabetes Endocrinol. **8**(4), 337–47 (2020)
4. Xu, Z., Escalera, S., Pavão, A., et al. Codabench: flexible, easy-to-use, and reproducible meta-benchmark platform. Patterns **3**(7) (2022)
5. Chen, P.N., Lee, C.C., Liang, C.M., et al.: General deep learning model for detecting diabetic retinopathy. BMC Bioinform. **22**(Suppl 5), 84 (2021)
6. Abràmoff, M.D., Lavin, P.T., Birch, M., et al.: Pivotal trial of an autonomous AI-based diagnostic system for detection of diabetic retinopathy in primary care offices. NPJ Digit. Med. **1**, 39 (2018)
7. Ting, D., Peng, L., Varadarajan, A.V., et al.: Deep learning in ophthalmology: the technical and clinical considerations. Progress Retinal Eye Res. **72**, 100759 (2019)

8. Liu, R., et al.: DeepDRiD: diabetic retinopathy-grading and image quality estimation challenge. Patterns **3**(6), 100512 (2022)
9. He, K., Zhang, X., Ren, S., Sun, J.: Deep residual learning for image recognition. In: Proceedings of the IEEE Conference on Computer Vision and Pattern Recognition, pp. 770–778 (2016)
10. Tan, M.: Efficientnet: rethinking model scaling for convolutional neural networks. arXiv preprint arXiv:1905.11946 (2019)

Automatic Prediction of Center-Involved Diabetic Macular Edema Using Ultra-Widefield Color Fundus Photography

Philippe Zhang[1,2,3](✉), Yihao Li[5], Weili Jiang[4], Jing Zhang[1,2], Sarah Matta[1,2], Yubo Tan[6], Hui Lin[7], Haoshen Wang[8], Jiangtian Pan[9], Hui Xu[10], Laurent Borderie[3], Alexandre Le Guilcher[3], Béatrice Cochener[1], Chubin Ou[11], Gwenolé Quellec[1], and Mathieu Lamard[1,2]

[1] LaTIM UMR 1101, Inserm, Brest, France
[2] University of Western Brittany, Brest, France
pzhang.wj88@gmail.com
[3] Evolucare Technologies, Villers-Bretonneux, France
[4] United Imaging Healthcare, Shanghai, China
[5] College of Computer Science, Sichuan University, Chengdu, China
[6] University of Electronic Science and Technology of China, Chengdu, China
[7] Northwestern University, Evanston, IL, USA
[8] ShanghaiTech University, Shanghai, China
[9] Wuhan University, Wuhan, China
[10] MD Anderson Cancer Center, Houston, USA
[11] Guangdong General Hospital, Guangzhou, Guangdong, China

Abstract. Centre-involved Diabetic Macular Edema (ci-DME) is a major cause of vision impairment and can arise at any stage of diabetic retinopathy (DR), with increased prevalence as DR progresses. Early detection is critical for preventing vision loss. Recent advancements in imaging techniques, such as ultra-wide-field color fundus photography (UWF-CFP) and optical coherence tomography (OCT), coupled with the power of deep learning (DL), now enable more precise detection, classification, and grading of DME. The **Device-Independent Diabetic Macular Edema Onset Prediction (DIAMOND) Challenge** aims to develop DL models capable of predicting whether a patient will develop center-involved diabetic macular edema (ci-DME) within a year using UWF-CFP images, while preventing participants from direct data access to ensure model generalizability in real-world settings. For this challenge, we implemented advanced preprocessing techniques to mitigate out-of-domain data issues and explored multiple DL architectures. Our methodology has shown great promise, positioning our team within the **Top 3** of the competition. However, due to the affiliation of some team members with the challenge organizers, we are ineligible for the prize.

Keywords: Diabetic Macular Edema · Ultra-wide-field Color Fundus Photography · Deep Learning · Prediction · Domain Generalization · Blind Challenge

1 Introduction

Center-involved diabetic macular edema (ci-DME) is a major cause of vision loss [6]. DME can develop at any stage of diabetic retinopathy (DR) and becomes increasingly prevalent as DR progresses to more severe stages. When DME affects or threatens the fovea, it is more likely to cause blurred vision and metamorphopsia. According to the Early Treatment Diabetic Retinopathy Study, when DME involves or threatens the fovea, the risk of moderate visual loss over a 3-year period is 24% without treatment [1]. Thus, early diagnosis is critical to prevent further progression and vision loss.

New imaging techniques, such as ultra-wide-field color fundus photography (UWF-CFP), optical coherence tomography (OCT), and OCT angiography, are already available and generate a wealth of valuable data [4]. These techniques, when combined with cutting-edge deep learning (DL) approaches, have significantly improved efficiency in various medical imaging applications such as retinal disease detection [2,3,5,8,9].

Among these imaging modalities, UWF-CFP stands out for its ability to capture more than 80% of the retina in a single high-resolution image, including peripheral lesions that can signal early signs of DME. This capability makes UWF-CFP particularly valuable for early detection and prediction tasks, as it allows for a comprehensive assessment of the retina, which is crucial for identifying subtle indicators of disease progression.

The challenge Device-Independent Diabetic Macular Edema Onset Prediction (DIAMOND) builds on this potential by aiming to develop and evaluate DL models that predict if a patient will develop ci-DME within a year, using UWF-CFP images exclusively. Unlike conventional challenges, participants in DIAMOND do not have direct access to datasets for training, validation, or testing. Instead, they submit their code, which is executed by the organizing committee. This setup encourages the development of highly generalizable models while addressing real-world constraints such as data privacy, ethical considerations, and logistical limitations.

2 Materials and Methods

2.1 Dataset

In this challenge, the dataset consists of UWF-CFP images captured using different devices across diverse demographic populations. These images are obtained from three distinct devices: **CLARUS**, **OPTOS**, and **EIDON**, as illustrated in Fig. 1.

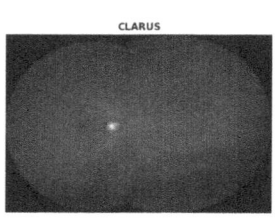 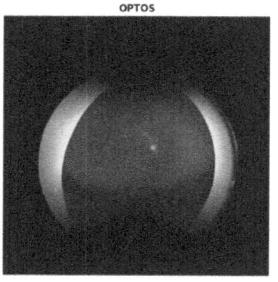 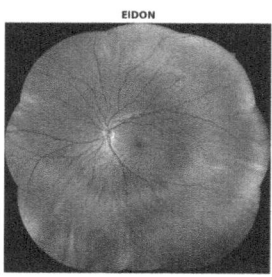

Fig. 1. Sample images from each domain. CLARUS and OPTOS images are taken from the sample data provided by the challenge organizers, while the EIDON image is sourced from the manufacturer's website.

Table 1. Domains Distributions of Challenge Dataset

Split	D1	D2	D3	D4
Train	✓	✓		
Validation	✓	✓		
Test	✓	✓	✓	✓

Based on sample data provided by the challenge organizers, which includes one CLARUS and one OPTOS image for each eye, CLARUS images have resolutions ranging from 4920 × 4234 to 6252 × 4300 pixels, while OPTOS images consistently have a resolution of 4000 × 4000 pixels. Although no EIDON images were included in the provided sample data, manufacturer specifications[1] indicate that these images typically have a resolution of 4608 × 3288 pixels. This variation in image resolutions, combined with differences in device-specific imaging techniques, highlights the need for rigorous preprocessing to harmonize data and ensure consistency in analysis.

The dataset is divided into three subsets: **Train**, **Validation**, and **Test**, with the distribution outlined in Table 1. More specifically, the dataset comprises the following domains:

- **D1**: CLARUS images from a European population,
- **D2**: OPTOS images from a European population,
- **D3**: OPTOS images from an African population,
- **D4**: EIDON images from an African population.

The **Train** and **Validation** sets include images exclusively from the **D1** and **D2** domains, covering CLARUS and OPTOS devices but limited to a European demographic. In contrast, the **Test** set includes images from all four domains (**D1**, **D2**, **D3**, and **D4**), introducing both modality variation (with three different devices) and demographic diversity (with both European and African

---

[1] https://edc-lamy.fr/ultra-grand-champ.

populations). This setup enables the challenge to rigorously evaluate the ability of models to generalize across different imaging technologies and demographic shifts.

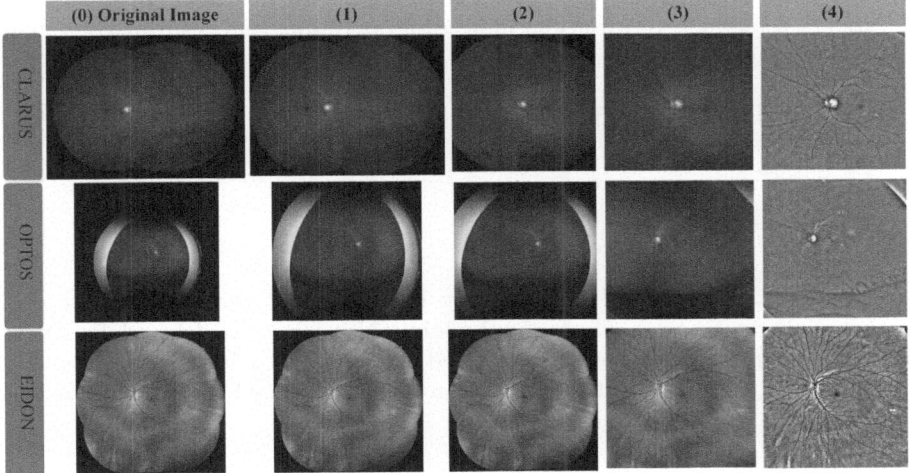

Fig. 2. Data Processing

2.2 Data Processing

As shown in Fig. 2, we perform our data preprocessing on retinal images captured from three different imaging devices (CLARUS, OPTOS, and EIDON). The preprocessing steps, applied consistently across all modalities, are as follows:

- **Original Image (Column 0)**: The raw, unprocessed images as captured by the respective devices. These images often contain varying levels of background noise, black borders, and inconsistent resolutions.
- **Removing Black Borders (Column 1)**: The first step is to remove any black borders surrounding the retinal images, if present (note that EIDON images do not have black borders). This focuses on the relevant retinal area, eliminating unnecessary background elements, and prepares the image for further processing.
- **Square Center Crop (Column 2)**: After border removal, the images are center-cropped into a square format, using the smaller dimension between width and height. This ensures uniform dimensions across samples, which facilitates consistent training and avoids distortions from resizing.
- **Resizing and Central Crop (Column 3)**: The images are then resized to a resolution of 1024×1024 pixels. From this resized image, a central region of 800×800 pixels is cropped. This maintains a standardized input size while focusing on the central retinal features critical for model performance.

- **Color Normalization (Column 4)**: Lastly, color normalization is applied to standardize the color distribution across all images. We adopt the approach proposed by Zhang et al. [7], which includes Gaussian blurring, local mean subtraction, amplification and offset adjustment, and channel reintegration. This method effectively reduces device-specific color biases, enhancing the model's robustness to variations in imaging conditions and ensuring consistent image quality across the dataset.

These preprocessing steps ensure consistency across all imaging modalities and demographic populations, improving the model's ability to generalize across diverse datasets.

2.3 Model Architectures

For this task, we explored two primary deep learning architectures for predicting diabetic macular edema: an **EfficientNet**-based model and **ResNet-50**. Both architectures have demonstrated strong performance in image classification tasks, but offer distinct advantages that we aimed to leverage for this challenge.

EfficientNet. We employed the Efficient-Net network to accurately and automatically predict central diabetes (see Fig. 3). EfficientNet is a convolutional neural network architecture introduced by Google's research team, designed to address the challenge of scaling models efficiently while maintaining performance. Unlike traditional models that typically scale along a single dimension-such as increasing depth, width, or input resolution-EfficientNet proposes a compound scaling method that balances all three dimensions simultaneously. The base model, EfficientNet-B0, was created using Neural Architecture Search (NAS) to optimize the trade-off between accuracy and efficiency. From this foundation, larger models (EfficientNet-B1 to B7) are systematically scaled up by increasing depth, width, and resolution, achieving significant improvements in performance while minimizing computational cost. EfficientNet also incorporates Squeeze-and-Excitation (SE) blocks, which enhance feature extraction by applying channel-wise attention, and MBConv layers, which reduce computational overhead using depthwise separable convolutions. Compared to traditional CNN architectures like ResNet or DenseNet, EfficientNet offers superior accuracy on benchmark datasets such as ImageNet while using fewer parameters and computational resources.

ResNet-50. We also employed the ResNet-50 network to accurately and automatically predict central diabetes (see Fig. 4). ResNet-50 contains 50 layers and is notable for introducing the concept of residual learning, which addresses the challenge of training very deep networks by using skip connections (or shortcuts) to bypass certain layers. These skip connections allow the model to learn identity mappings more easily, thus preventing the degradation of accuracy as

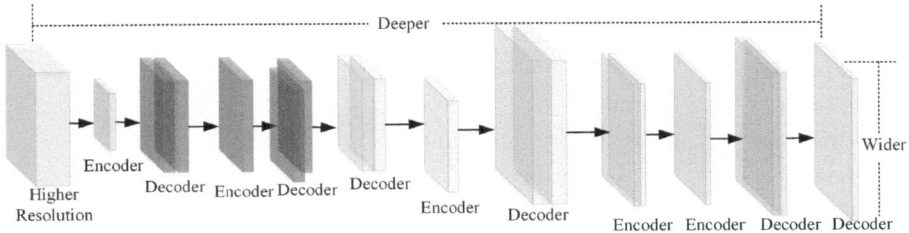

Fig. 3. Efficient-Net Structure

the network depth increases, which was a common issue with deeper networks before ResNet. The core idea behind ResNet-50's architecture is the use of residual blocks, where the input is passed forward and added to the output of a few stacked layers. This architecture allows gradients to flow more easily during backpropagation, solving the vanishing gradient problem and enabling efficient training of deep networks. ResNet-50 is composed of a series of convolutional layers, batch normalization, ReLU activation, and max-pooling, followed by fully connected layers.

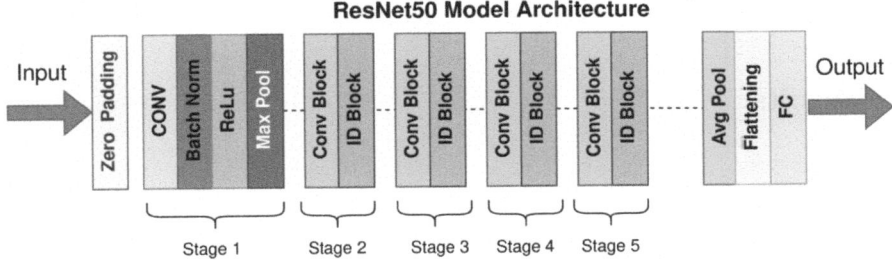

Fig. 4. Resnet-50 Structure

2.4 Implementation Details

The implementation details of the models used in our experiments are summarized in Table 4. All models were implemented using the `timm` library and pre-trained on the ImageNet dataset. After preprocessing, all UWF-CFP images were resized and center-cropped to a resolution of 800 × 800 pixels. For model development, the training and validation sets were predefined by the challenge organizers, ensuring consistent evaluation protocols across participants. Data augmentation techniques, including `RandomHorizontalFlip`, `RandomVerticalFlip`, `RandomRotation`, `RandomPerspective`, `RandomAdjustSharpness`, and `ColorJitter`, were applied to enhance model robustness against variations in imaging

Table 2. Implementation details used in experiments.

Input size	800 × 800 pixels
Library	timm
Pretrained weights	see Table 3, and Table 4
Loss	CrossEntropyLoss
Optimizer	AdamW
Learning rate	1e−4 (w/o scheduler)
Augmentation	RandomHorizontalFlip, RandomVerticalFlip, RandomRotation, RandomPerspective, RandomAdjustSharpness, ColorJitter
Batch size	16
Epochs/Times	400/30 h
Train/Val split	Predefined split by the challenge organizers
Metric	**AUC**, F1, ECE

conditions. The best-performing model was selected based on the highest AUC score achieved on the validation set. All training and evaluations were conducted on a Tesla V100 GPU with 32 GB of memory, enabling efficient processing of the high-resolution images and DL models (Table 2).

3 Results

Table 3. Preliminary Model Performance on the DIAMOND Challenge

Backbone Model	Pretrained Weight	AUC	F1 Score	ECE
`tf_efficientnetv2_l`	ImageNet	0.5710	0.1759	0.2068
`tf_efficientnetv2_b2`	ImageNet	0.6172	0.0	0.0767
`tf_efficientnet_b2`	ImageNet	0.5791	0.0	0.0766
`resnet50`	ImageNet	0.6773	0.0	0.0162

The preliminary results from the DIAMOND Challenge, as summarized in Table 3, compare the performance of various backbone models pre-trained on ImageNet across several evaluation metrics. Among the models, ResNet-50 demonstrated the best performance with the highest AUC of 0.6773 and the lowest ECE (Expected Calibration Error) of 0.0162, indicating both strong discriminatory power and good calibration. However, its F1 score is recorded as 0.0, likely due to poor thresholding or class imbalance, suggesting a poor balance between precision and recall. The tf_efficientnetv2_b2 model follows with a moderate AUC of 0.6172 and an ECE of 0.0767, but it also has an F1 score of 0.0, indicating possible issues with class imbalance or thresholding strategies in the model's predictions. Similarly, the tf_efficientnet_b2 and tf_efficientnetv2_l models achieve AUCs of 0.5791 and 0.5710, respectively. However, both also have an F1 score of 0.0, with higher ECE values (0.0766 and 0.2068), indicating worse calibration compared to ResNet-50.

Table 4. Performance of Pre-trained Models on UWF4DR Dataset

Backbone Model	Pretrained Weight	AUC	F1 Score	ECE
tf_efficientnetv2_b2	UWF4DR	0.6211	0.0	0.0767
tf_efficientnet_b2	UWF4DR	0.5928	0.0	0.0767

We present in Table 4 the performance of models pre-trained on the UWF4DR dataset. The UWF4DR[2] dataset was introduced in another competition held at MICCAI 2024, which focused on the identification of DR and DME using UWF-CFP. We leveraged the pre-trained weights from models developed for the UWF4DR competition and fine-tuned them for the DIAMOND challenge. Specifically, we used the best-performing models from the UWF4DR competition, which achieved the highest AUC for DME detection during the experimental phase.

Due to the UWF4DR competition's specific evaluation, only lightweight models (tf_efficientnetv2_b2 and tf_efficientnet_b2) were trained because of their faster inference speed and strong performance, which were key metrics evaluated for ranking in this challenge.

As demonstrated in Table 4, after incorporating pre-training on the UWF4DR dataset for DME identification, both tf_efficientnetv2_b2 and tf_efficientnet_b2 showed improved AUC compared to the results in Table 3. However, the F1 score remains at 0.0, highlighting a persistent challenge in achieving a balance between precision and recall in classification.

4 Conclusions and Discussion

In this study, we investigated the use of deep learning models, specifically EfficientNet and ResNet-50, for the automatic prediction of ci-DME from UWF-CFP. As ci-DME is a leading cause of vision loss among diabetic patients, early detection is essential for preventing disease progression. By leveraging advanced imaging techniques and DL, there is significant potential for enhancing early diagnosis.

Our results demonstrate that ResNet-50 consistently outperforms the EfficientNet models, achieving the highest AUC (0.6773) and the lowest ECE (0.0162) in the DIAMOND Challenge. However, the F1 score for all models

---
[2] https://codalab.lisn.upsaclay.fr/competitions/18605.

remained at 0.0, indicating persistent challenges in balancing precision and recall. Pre-training on the UWF4DR dataset led to marginal improvements in AUC for EfficientNet models, though the F1 score issues persisted.

The UWF4DR Challenge focused on detecting DR and DME using UWF-CFP captured exclusively with the OPTOS device [7]. On the other hand, the DIAMOND Challenge shifts the focus to predicting ci-DME onset within a year, using UWF-CFP captured from multiple devices (CLARUS, OPTOS, and EIDON) to model temporal progression. These differences highlight the complementary roles of the two challenges. UWF4DR provides a robust framework for learning spatial features indicative of DR and DME, while DIAMOND introduces additional complexity by requiring generalization across devices and prediction of temporal disease progression.

Building on this complementarity, we hypothesize that pre-training on the UWF4DR dataset offers added value by initializing models with robust weights that capture essential retinal features, particularly those linked to DR and DME. Fine-tuning these weights for ci-DME prediction in DIAMOND allows the model to adapt to the task's specific requirements more effectively. While submission constraints prevented us from testing the impact of UWF4DR pre-training on ResNet-50, its strong performance in DIAMOND suggests that incorporating this step could further enhance its effectiveness. This approach could improve both AUC and F1 scores by enabling the model to capture relevant retinal features more effectively.

Future work will focus on evaluating this pre-training strategy for ResNet-50, integrating UWF-CFP dataset such as UWF4DR, and exploring advanced techniques to address current limitations. Specifically, strategies like model ensembling and improved preprocessing techniques could help tackle challenges such as class imbalance and domain variability, thereby improving robustness and generalizability.

Disclosure of Interests. The authors have no competing interests.

References

1. Aiello, L.M., Ferris, F.L.: Photocoagulation for diabetic macular edema. Arch. Ophthalmol. **105**(9), 1163 (1987)
2. El Habib Daho, M., et al.: Improved automatic diabetic retinopathy severity classification using deep multimodal fusion of UWF-CFP and OCTA images. In: Antony, B., Chen, H., Fang, H., Fu, H., Lee, C.S., Zheng, Y. (eds.) Ophthalmic Medical Image Analysis, pp. 11–20. Springer, Cham (2023). https://doi.org/10.1007/978-3-031-44013-7_2
3. El Habib Daho, M., et al.: DISCOVER: 2-D multiview summarization of optical coherence tomography angiography for automatic diabetic retinopathy diagnosis. Artif. Intell. Med. **149**, 102803 (2024). https://doi.org/10.1016/j.artmed.2024.102803
4. Kalra, G., et al.: Recent advances in wide field and ultrawide field optical coherence tomography angiography in retinochoroidal pathologies. Expert Rev. Med. Devices **18**(4), 375–386 (2021)

5. Li, Y., et al.: Automated detection of myopic maculopathy in MMAC 2023: achievements in classification, segmentation, and spherical equivalent prediction. In: International Conference on Medical Image Computing and Computer-Assisted Intervention, pp. 1–17. Springer, Cham (2023). https://doi.org/10.1007/978-3-031-54857-4_1
6. Varadarajan, A.V., et al.: Predicting optical coherence tomography-derived diabetic macular edema grades from fundus photographs using deep learning. Nat. Commun. **11**(1), 130 (2020)
7. Zhang, P., Conze, P.H., Lamard, M., Quellec, G., El Habib Daho, M : Deep learning-based detection of referable diabetic retinopathy and macular edema using ultra-widefield fundus imaging (2024). https://arxiv.org/abs/2409.12854
8. Zhang, P., et al.: Detection and classification of glaucoma in the justraigs challenge: achievements in binary and multilabel classification. In: 2024 IEEE International Symposium on Biomedical Imaging (ISBI), pp. 1–4. IEEE (2024). https://doi.org/10.1109/ISBI56570.2024.10635113
9. Zhou, S.K., et al.: A review of deep learning in medical imaging: imaging traits, technology trends, case studies with progress highlights, and future promises. Proc. IEEE **109**(5), 820–838 (2021)

A Device-Agnostic Deep Learning Approach for Predicting Ci-DME Onset Using UWF-CFP Images

Amin Khouani[1]($^{\boxtimes}$) and Ihsane Mekki[2]

[1] Higher School of Computer Science, Algiers, Algeria
a_khouani@esi.dz
[2] Abou Bakr Belkaid University, Tlemcen, Algeria

Abstract. Center-involved diabetic macular edema (ci-DME) is a sight-threatening complication of diabetes. Early detection is crucial for preventing vision loss. This paper proposes a deep learning-based method using Ultra-Wide Field Color Fundus Photography (UWF-CFP) images to predict ci-DME development within one year. The challenge lies in handling images from different devices (Optos, Clarus, Eidon) without access to training data. Our approach employs a two-step model: first, a device detection model identifies the imaging device, followed by device-specific classification models for ci-DME prediction. The proposed method achieves an AUC score of 0.9860 on a simulated validation set. However, further refinement is needed, as evidenced by the low F1 score and calibration error. We discuss potential improvements using transformers and foundation models like RetFound for enhanced ci-DME prediction.

Keywords: Diabetic Macular Edema (DME) · Deep Learning in Ophthalmology · Ultra-Wide Field Color Fundus Photography (UWF-CFP)

1 Introduction

Diabetic Macular Edema (DME) is one of the most frequent causes of vision loss in people with diabetes, affecting approximately 21 million individuals worldwide [1]. DME occurs when blood vessels in the retina, weakened by prolonged exposure to high blood glucose levels, leak fluid into the macula, causing swelling and impaired vision. A specific form of DME, known as center-involved diabetic macular edema (ci-DME), is particularly harmful as it affects the central region of the retina responsible for detailed vision, critical for reading and recognizing faces. Untreated ci-DME can lead to irreversible vision loss, significantly reducing quality of life [2].

Early detection and timely treatment of ci-DME are crucial for preventing vision deterioration [3]. Current standard diagnostic tools involve optical coherence tomography (OCT) and clinical examination; however, these methods can be resource-intensive and limited by geographical access in certain regions. A non-invasive and efficient alternative is fundus photography, particularly Ultra-Wide Field Color Fundus Photography (UWF-CFP), which provides a broader view of the retina, making it a valuable tool for early detection of diabetic retinopathy (DR) and associated complications such as ci-DME [4].

Despite advances in imaging techniques, the task of predicting ci-DME onset from UWF-CFP images within a year remains a challenging problem due to several factors. First, the disease progresses gradually, and early retinal changes indicative of ci-DME can be subtle, making detection difficult even for trained specialists. Additionally, the quality of UWF-CFP images can vary significantly between imaging devices, such as Optos, Clarus, and Eidon, each of which captures the retina differently [5]. For example, Optos devices offer a 200-degree view of the retina with greater peripheral visualization, while Clarus systems capture high-resolution images closer to conventional fundus photography [6]. Eidon devices provide high-definition, wide-field imaging that closely resembles the quality of Clarus but includes additional diagnostic capabilities.

The DIAMOND Challenge, which this paper focuses on, was designed to address this predictive task. The challenge requires participants to develop models capable of predicting the development of ci-DME within a year using only UWF-CFP images. Participants are restricted from accessing training, validation, or test datasets directly and instead submit their code to a secure cloud-based system where the models are evaluated. This setup reflects real-world constraints such as data privacy and the need for generalizable models, making it an ideal platform for advancing research in the prediction of ci-DME.

The challenge dataset includes images from multiple devices, including Optos, Clarus, and Eidon, with their unique imaging characteristics. For example:

Optos Images: Provide an ultra-wide 200-degree view of the retina, capturing both the central and peripheral regions. These images tend to have more peripheral detail but may suffer from lower resolution in the central retina.

Clarus Images: Offer higher resolution images focusing more on the central retina, with an approximately 133-degree field of view, capturing more detailed macular structures.

Eidon Images: Provide high-definition imaging similar to Clarus but with enhanced color and diagnostic accuracy. The task requires the model to not only predict the onset of ci-DME but also generalize across different imaging devices, adding complexity to the challenge.

Artificial Intelligence (AI) has revolutionized many fields of medicine, including ophthalmology, by enhancing the accuracy and efficiency of disease detection and prediction [7–9]. AI, particularly deep learning (DL) and convolutional neural networks (CNNs), has been applied to retinal images for diagnosing diabetic retinopathy (DR), DME, glaucoma, and age-related macular degeneration (AMD), showing comparable or even superior performance to human experts in some cases [10–13].

AI's ability to analyze large datasets and learn from subtle patterns in retinal images makes it particularly suited for predicting ci-DME. In the context of diabetic retinopathy and DME, AI models trained on fundus images can detect microaneurysms, exudates, and other retinal abnormalities indicative of disease progression [14]. These models offer the potential to automate the screening process, making early detection of conditions like ci-DME more accessible, especially in areas where specialized ophthalmologists may not be readily available.

1.1 Related Works

Numerous studies have demonstrated the effectiveness of AI in this domain. Below are some key works in the literature, showcasing AI's application in ophthalmology and ci-DME prediction:

Gulshan et al. (2016) [15]: In a landmark study, this research developed a deep learning model for detecting diabetic retinopathy (DR) and diabetic macular edema (DME) using retinal fundus images. The model achieved a sensitivity of 90.3% and a specificity of 98.1%, demonstrating that AI could perform as well as expert ophthalmologists. This study paved the way for AI-based screening tools for DR and DME.

Abràmoff et al. (2018) [16] This study introduced an AI system for autonomous detection of DR. The model was the first FDA-approved AI diagnostic system for ophthalmology, achieving a sensitivity of 87% and specificity of 90%. The system significantly reduced the workload for clinicians by identifying patients needing referral.

Li et al. (2019) [17] Developed a hybrid AI model combining deep learning and clinical data to predict the presence of DME from retinal images. The model achieved an AUC of 0.90 for detecting DME, highlighting the importance of incorporating clinical information alongside imaging data for more accurate predictions.

Schmidt-Erfurth et al. (2018) [18]: Proposed a deep learning model for predicting the progression of diabetic retinopathy based on longitudinal optical coherence tomography (OCT) data. Although focused on OCT, this work is relevant for understanding how AI can predict disease progression over time, similar to the prediction of ci-DME.

Bellemo et al. (2019) [19]: Investigated the feasibility of using AI to detect DME and DR in primary care settings. The study found that AI models could significantly improve the accuracy of DME detection in non-specialist settings, with a sensitivity of 97.6% and specificity of 95.7%, indicating that AI can help bridge the gap in areas lacking ophthalmologists.

Gargeya and Leng (2017) [20]: Developed a CNN model for automated detection of DR and its complications, including DME. The model achieved an AUC of 0.94 for DR and 0.85 for DME detection, suggesting that CNNs can effectively differentiate between normal and pathological retinal images.

Keel et al. (2018) [21]: This study demonstrated that a deep learning system could accurately detect referable DR, including DME, from fundus images with a sensitivity of 92.0% and specificity of 93.7%, further establishing the role of AI in automated screening for retinal diseases.

Xie et al. (2020) [22]: Developed a novel AI-based system using generative adversarial networks (GANs) to improve the quality of retinal images captured under suboptimal conditions. This enhancement led to better diagnostic accuracy for detecting ci-DME, especially in resource-limited settings where image quality may be compromised.

These works collectively demonstrate the vast potential of AI in transforming the field of ophthalmology by providing fast, accurate, and scalable solutions for detecting and predicting conditions like ci-DME. However, as most studies focus on OCT or traditional fundus images, there is a growing need to extend these advancements to UWF-CFP images, given their wide-field perspective of the retina. This research aims to fill that gap by developing a predictive model capable of forecasting ci-DME development using UWF-CFP images, while accounting for the complexities of multiple imaging devices.

2 Materials and Methods

2.1 Dataset

The dataset for this study was derived from the DIAMOND Challenge and includes UWF-CFP images captured using devices such as Optos, Clarus, and Eidon. The dataset forms part of the broader EVIRED project, which spans 14 French hospitals, aiming to predict ci-DME development and related complications. The challenge dataset does not provide direct access to training, validation, or test sets. Instead, participants submit their models for testing on a cloud-based platform. Example UWF-CFP images from each device are presented in Figure Fig. 1. The initial exploration of the dataset using a baseline ResNet-50 model achieved an AUC of 0.73, highlighting the feasibility of predicting ci-DME using UWF-CFP images.

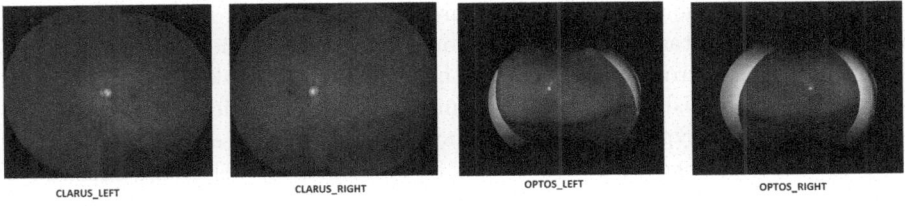

Fig. 1. Example UWF-CFP images from each device

2.2 Training Dataset and Initial Insights

The EVIRED dataset contains thousands of UWF-CFP images annotated for ci-DME. In this study, we focus on two types of imaging devices: Optos and Clarus. The challenge organizers used data from 14 French hospitals as part of the EVIRED project (https://evired.org/). The primary challenge was designing a model capable of differentiating between images from these devices and applying specialized classifiers accordingly.

2.3 Methodological Challenge

The DIAMOND Challenge presents a unique approach by not allowing direct access to the training and testing datasets. Participants submit their code, which is then executed on a dedicated cloud cluster. This approach fosters the development of generalizable models, reflecting real-world scenarios with data privacy and logistical limitations.

2.4 Proposed Models

In this study, we propose a two-stage predictive modeling approach for the development of ci-DME using UWF-CFP images. The first stage involves device detection, which classifies the imaging device used to capture the fundus image (Optos, Clarus, or Eidon). The second stage involves ci-DME classification, with separate models trained specifically for the detected device type (Optos or Clarus). This approach helps mitigate the variability in image characteristics between devices and ensures that the model's predictions are device-specific, leading to better performance.

Device Detection. The first model is designed to detect the device that captured the image, which is crucial given the distinct characteristics of images from different devices (Optos, Clarus, and Eidon). The device detection model is based on the EfficientNetV2 architecture, chosen for its strong performance in image classification tasks and computational efficiency. EfficientNetV2 excels at extracting features from high-resolution medical images, making it particularly suitable for distinguishing between the imaging characteristics of these devices.

Once the device type is predicted, the system dynamically switches to the corresponding device-specific model: either the emc_optos model or the emc_clarus model, which are tailored for ci-DME prediction. This dynamic switching ensures that the downstream classification task benefits from device-specific feature extraction, thereby improving overall model performance.

Classification Models. For the ci-DME classification task, we designed two specialized models: emc_optos and emc_clarus, each trained on data from their respective imaging device.

Table 1. Hyperparameters

Parameter	Value	Justification
Backbone	tf_efficientnetv2_xl	Chosen for its high accuracy and efficiency on image classification tasks. EfficientNetV2 has been shown to outperform ResNet-50 on various benchmarks.
Image size	448 × 448	Balances resolution with computational efficiency; high enough to capture critical retinal features while avoiding memory overhead.
Optimizer	AdamW	AdamW improves training stability and generalization by decoupling weight decay from the optimization steps.
Loss function	CrossEntropyLoss	Standard for classification tasks, appropriate for multi-class prediction.
Learning rate	1e−4	A low learning rate ensures stable convergence while avoiding large oscillations in the loss function.
Batch size	2	Small batch size due to memory constraints, particularly for large images and complex models like EfficientNetV2.
Scheduler	OneCycleLR	Dynamically adjusts learning rate to speed up convergence while preventing overfitting, especially in later epochs.
Weight decay	1e−4	Prevents overfitting by penalizing large weights, encouraging simpler model behavior.
Epochs	100	Allows the model sufficient time to learn the complex patterns in UWF-CFP images while avoiding overtraining.

The base architecture for these models is tf_efficientnetv2_xl [23], a powerful and efficient convolutional neural network (CNN) architecture known for its high accuracy and computational efficiency. We chose this architecture to ensure that the models could handle the variability and complexity of UWF-CFP images while remaining computationally feasible for large-scale deployment.

Each model was trained with specific hyperparameters, as summarized in Table 1.

We selected these hyperparameters based on a combination of theoretical insights and empirical experimentation.

The tf_efficientnetv2_xl backbone was chosen for its strong performance in medical image classification tasks, given its efficiency and scalability. The image size of 448 × 448 was selected to ensure sufficient detail was captured in each retinal image, particularly important for identifying subtle features associated with ci-DME.

The AdamW optimizer was chosen over traditional Adam due to its improved handling of weight decay, which is important for models trained on medical images, where overfitting is a concern. The learning rate of 1e−4 ensures slow and steady convergence, while the OneCycleLR scheduler dynamically adjusts the learning rate, boosting convergence speed while preventing overfitting in the later epochs.

The CrossEntropyLoss function is standard for multi-class classification tasks, which suits the device and ci-DME classification problems. The batch size of 2 was set due to the computationally expensive nature of training large CNNs on high-resolution images, and 100 epochs provided enough iterations for the model to learn while avoiding overfitting.

2.5 Key Highlights in Methodology

Device Switching. One of the most critical features of our proposed approach is device switching. The DiamondModelFinal architecture dynamically chooses which sub-model to use for ci-DME classification based on the predicted device type. This ensures that the system applies the most suitable model for the given image, reducing the risk of misclassifications caused by variations between devices.

For instance, once the model_device identifies an image as captured by Optos, the system automatically applies the emc_optos model, which was trained specifically for Optos images. This dynamic switching enables the model to handle images from multiple sources in a unified framework.

Pretrained Models. Our models are initialized with pretrained weights to speed up training and enhance model performance. Pretrained weights allow the models to start from a strong baseline, reducing the time needed to converge and improving the model's ability to generalize.

Data Augmentation. We used various data augmentation techniques to improve the model's generalization capabilities. These included random flips, rotations, and sharpness adjustments, which simulate the variations in real-world UWF-CFP images. These augmentations were critical for creating a robust model that could handle diverse imaging conditions.

Metrics. We evaluated the models using multiple metrics to provide a comprehensive understanding of their performance. These included:

– AUC (Area Under the Curve): Measures the model's ability to distinguish between ci-DME and non-ci-DME cases.
– F1 Score: Balances precision and recall, which is particularly important for imbalanced datasets like this one.

– Calibration Error (ECE): Assesses how well the model's predicted probabilities are calibrated, ensuring that confidence scores align with actual outcomes. This is crucial in medical imaging tasks where misclassifications can have serious consequences.

The following section presents the Results and Discussion, where we report the performance metrics achieved by our models and analyze their strengths and limitations. We discuss the accuracy, AUC, F1 score, and calibration error (ECE) in detail, and explore potential improvements for future work.

3 Results and Discussion

The proposed models were evaluated on the DIAMOND Challenge dataset, achieving an AUC score of 0.9860 on a simulated validation set, indicating strong performance in ci-DME prediction. However, the F1 score of 0.5244 and a calibration error of 0.0767 indicate that the model may face challenges related to class imbalance and calibration issues.

The low F1 score could be attributed to the limited number of positive samples (ci-DME cases) in the dataset. Addressing this imbalance might require techniques like oversampling or undersampling. Additionally, exploring different loss functions or class weighting strategies could help improve the model's sensitivity to positive cases.

The calibration error indicates that the model's predicted probabilities may not accurately reflect the true probabilities of ci-DME. This can impact decision-making, as clinicians may rely on these probabilities to assess the risk of disease progression. Techniques like temperature scaling or Platt scaling can be used to calibrate the model's outputs.

4 Conclusion

This paper presents a deep learning approach for predicting ci-DME onset using UWF-CFP images. The proposed method leverages a two-step architecture for device-agnostic prediction. Which first detects the imaging device and then applies a device-specific classification model, addresses the challenge of working with images from multiple sources.

While our results show some promise, particularly in terms of the overall score and calibration, the low AUC and F1 score indicate that significant improvements are needed. Future work could explore several avenues to enhance the model's performance:

Larger Datasets: Access to larger and more diverse datasets would enable the training of more robust models.

Advanced Architectures: Incorporating more complex deep learning architectures, such as transformers, could enhance feature extraction and prediction capabilities.

Foundation Models: Leveraging pre-trained foundation models specifically designed for medical image analysis could provide a strong starting point for ci-DME prediction.

Multimodal Approaches: Combining UWF-CFP with other imaging modalities, such as OCT, could improve prediction accuracy.

Explainability and Interpretability: Focus on enhancing the explainability and interpretability of our model by integrating techniques such as Grad-CAM, which can provide visual explanations for the regions of interest contributing to the model's predictions. These efforts aim to align the model's outputs with clinical understanding, facilitating trust and usability in medical settings while supporting transparent decision-making processes.

Optos Image Preprocessing: Incorporating a pre-processing step for Optos images to intelligently crop the informative regions, reducing the impact of non-informative areas and optimizing the resolution for more effective feature extraction.

Class Imbalance: Addressing the class imbalance present in the dataset by implementing advanced techniques such as weighted sampling, class-balanced loss functions, and oversampling strategies to enhance model sensitivity to underrepresented cases.

Central Region Analysis: A comparative analysis of models trained on full retinal images versus those trained on cropped central regions will be explored to assess the predictive power of central biomarkers in ci-DME development.

By addressing these limitations and exploring future directions, we can advance the development of AI-powered tools for early ci-DME detection and improve the management of diabetic retinopathy.

References

1. World Health Organization: Global Report on Diabetes (2016)
2. Lang, G.E.: Diabetic macular edema. Ophthalmologica **227**(Suppl. 1), 21–29 (2012). https://doi.org/10.1159/000337156
3. Zhang, P., Conze, P.-H., Lamard, M., Quellec, G., El Habib Daho, M.: Deep learning-based detection of referable diabetic retinopathy and macular edema using ultra-widefield fundus imaging. arXiv preprint, arXiv:2409.12854 (2024). https://arxiv.org/abs/2409.12854
4. Singer, M., et al.: Ultra-widefield imaging of the peripheral retina in diabetic retinopathy: a review. Clin. Ophthalmol. **10**, 2035–2045 (2016)
5. El Habib Daho, M., et al.: Cross-device AI fusion: enhancing diabetic retinopathy diagnosis with combined clarus and optos images. Invest. Ophthalmol. Vis. Sci. **65**(7), 5630 (2024)
6. El Habib Daho, M., et al.: Performance of two ultra-widefield retinal imaging systems for the automatic diagnosis of diabetic retinopathy. Investigative Ophthalmol. Vis. Sci. **64**(8), 251–251 (2023)

7. Mahmoudi, S.A., Stassin, S., Daho, M., Lessage, X., Mahmoudi, S.: Explainable deep learning for Covid-19 detection using chest X-ray and CT-scan images. In: Garg, L., Chakraborty, C., Mahmoudi, S., Sohmen, V.S. (eds.) Healthcare Informatics for Fighting COVID-19 and Future Epidemics. EICC, pp. 311–336. Springer, Cham (2022). https://doi.org/10.1007/978-3-030-72752-9_16
8. Li, Y., et al.: Segmentation, classification, and quality assessment of UW-OCTA images for the diagnosis of diabetic retinopathy. In: Mitosis Domain Generalization and Diabetic Retinopathy Analysis, MIDOG DRAC 2022. Lecture Notes in Computer Science, vol. 13597. Springer, Cham (2023)
9. El Habib Daho, M., et al.: DISCOVER: 2-D multiview summarization of Optical Coherence Tomography Angiography for automatic diabetic retinopathy diagnosis. Artif. Intell. Med. **149**, 102803 (2024). https://doi.org/10.1016/j.artmed.2024.102803
10. Li, F., et al.: A deep-learning system predicts glaucoma incidence and progression using retinal photographs. J. Clin. Investigation **132**(11) (2022). https://doi.org/10.1172/JCI157968
11. Akter, N., Fletcher, J., Perry, S., et al.: Glaucoma diagnosis using multi-feature analysis and a deep learning technique. Sci. Rep. **12**(8064) (2022). https://doi.org/10.1038/s41598-022-12147-y
12. Zeghlache, R., et al.: Detection of diabetic retinopathy using longitudinal self-supervised learning. In: Antony, B., Fu, H., Lee, C.S., MacGillivray, T., Xu, Y., Zheng, Y. (eds.) Ophthalmic Medical Image Analysis, OMIA 2022. LNCS, vol. 13576, pp. 52–61. Springer, Cham (2022). https://doi.org/10.1007/978-3-031-16525-2_5
13. El Habib Daho, M., et al.: Improved automatic diabetic retinopathy severity classification using deep multimodal fusion of UWF-CFP and OCTA images. In: Ophthalmic Medical Image Analysis, OMIA 2023. LNCS, vol. 14096, Springer, Cham (2023). https://doi.org/10.1007/978-3-031-44013-7_2
14. Prathibha, S., Siddappaji: Advancing diabetic retinopathy diagnosis with fundus imaging: a comprehensive survey of computer-aided detection, grading and classification methods. Glob. Trans. **6**, 93–112 (2024). https://doi.org/10.1016/j.glt.2024.04.001
15. Gulshan, V., et al.: Development and validation of a deep learning algorithm for detection of diabetic retinopathy in retinal fundus photographs. JAMA **316**(22), 2402–2410 (2016)
16. Abràmoff, M.D., Lavin, P.T., Birch, M., Shah, N., Folk, J.C.: Pivotal trial of an autonomous AI-based diagnostic system for detection of diabetic retinopathy in primary care offices. NPJ Digit. Med. **1**(1), 39 (2018)
17. Li, F., Chen, H., Liu, Z., Zhang, X., Wu, Z.: Fully automated detection of retinal disorders by image-based deep learning. Graefes Arch. Clin. Exp. Ophthalmol. **257**, 495–505 (2019)
18. Schmidt-Erfurth, U., Sadeghipour, A., Gerendas, B.S., Waldstein, S.M., Bogunović, H.: Artificial intelligence in retina. Prog. Retin. Eye Res. **67**, 1–29 (2018)
19. Bellemo, V., et al.: Artificial intelligence using deep learning to screen for referable and vision-threatening diabetic retinopathy in Africa: a clinical validation study. Lancet Digit. Health **1**(1), e35–e44 (2019)
20. Gargeya, R., Leng, T.: Automated identification of diabetic retinopathy using deep learning. Ophthalmology **124**(7), 962–969 (2017)

21. Keel, S., et al.: Feasibility and patient acceptability of a novel artificial intelligence-based screening model for diabetic retinopathy at endocrinology outpatient services: a pilot study. Sci. Rep. **8**(1), 4330 (2018)
22. Xie, H., et al.: AMD-GAN: attention encoder and multi-branch structure based generative adversarial networks for fundus disease detection from scanning laser ophthalmoscopy images. Neural Netw. **132**, 477–490 (2020)
23. Tan, M., Le, Q.V.: EfficientNetV2: smaller models and faster training. CoRR, vol. abs/2104.00298 (2021). https://arxiv.org/abs/2104.00298

Onset Prediction of Center-Involved Diabetic Macular Edema on Ultra-Widefield Fundus Images Using Finetuned ResNet with Class Balanced Focal Loss

Yasine Lehmiani[1(✉)], Ayoub Elkhouzari[1], and Abdelhak Mahmoudi[1,2]

[1] LIMIARF, Faculty of Sciences, Mohammed V University in Rabat, Rabat, Morocco
yasine.lehmiani@um5r.ac.ma, abdelhak.mahmoudi@um5.ac.ma
[2] Ecole Normale Supérieure, Mohammed V University in Rabat, Rabat, Morocco

Abstract. This study presents an evaluation of three deep learning approaches submitted to the MICCAI's 2024 DIAMOND (Device-Independent diAbetic Macular edema ONset preDiction) challenge, aimed at predicting the onset of center-involved diabetic macular edema (ci-DME) within a year based on Ultra-Widefield Fundus Images. Participants didn't have access to the training, validation, or test datasets, making it necessary to develop models solely using a DeepDRiD synthetic dataset. We experimented with various deep learning architectures, including ResNet101, InceptionV3, and ResNet152, alongside different loss functions such as cross-entropy and class-balanced focal loss, complemented by data augmentation techniques. Our best-performing submission, using ResNet152, achieved an AUC of 0.798, an F1 score of 0.65, and an ECE of 0.11 on the DeepDRiD synthetic dataset.

1 Introduction

Diabetic Retinopathy (DR) [1] is a severe eye condition that affects individuals with diabetes, occurring when high blood sugar levels damage the blood vessels in the retina. Over time, these damaged vessels may swell, leak, or close off completely, which, if left untreated, can lead to significant vision loss. DR is classified into various stages, beginning with mild non-proliferative diabetic retinopathy (NPDR), where microaneurysms form, to severe NPDR, characterized by blocked blood vessels. The most advanced stage is proliferative diabetic retinopathy (PDR), where abnormal blood vessels grow on the surface of the retina, potentially causing severe vision impairment or blindness.

A major complication of DR is diabetic macular edema (DME) [2], which involves the accumulation of fluid in the macula, the part of the retina responsible for sharp, central vision. This fluid leakage leads to macular swelling, causing blurred or distorted vision. A specific type of DME, known as center-involved diabetic macular edema (ci-DME), occurs when retinal thickening involves the

© The Author(s), under exclusive license to Springer Nature Switzerland AG 2025
G. Quellec et al. (Eds.): DIAMOND 2024/MARIO 2024, LNCS 15503, pp. 41–53, 2025.
https://doi.org/10.1007/978-3-031-86651-7_4

central subfield zone of the macula. A patient is considered to be at risk of ci-DME in future visits if the DR label is at least moderate non-proliferative DR (NPDR). Otherwise, the condition is classified as non-ci-DME.

Detecting ci-DME can be performed from 3D optical coherence tomography (OCT) imaging using deep learning [3–5]. However, OCT is neither cost-effective nor widely accessible, particularly in resource-limited settings [6]. Recent study demonstrate that it is possible to train deep learning models to predict the presence of ci-DME with a sensitivity of 85% and a specificity of 80% using only 2D color fundus photography (CFP) [7]. Another study using CFP to detect DME demonstrated that, in a combined dataset from Australia, India, and Thailand, the deep learning CNN achieved 80% specificity and 81% sensitivity [8]. When compared to human experts, the proposed model had significantly higher specificity (P = 0.008) and non-inferior sensitivity (P < 0.001).

The DIAMOND challenge [9] aims to push this further by predicting the presence of ci-DME within a year using ultra-wide-field color fundus photography (UWF-CFP), an advanced form of CFP. The goal of this challenge is to develop methods that predict whether a patient will develop ci-DME within one year, based solely on 2D UWF-CFP images. A key difficulty lies in the fact that the train, validation and test data named EviRed dataset is unseen [10], due to privacy, ethical, and logistical considerations, and participants must rely on synthetic dataset generated from 2D UWF-CFP images from the Deep Diabetic Retinopathy Image Dataset (DeepDRiD) [11] available under the Creative Commons Attribution Share Alike 4.0 International license. In this synthetic dataset, the ci-DME labels were generated based on DR severity.

In this paper, we compared three approaches for predicting the presence of center-involved diabetic macular edema (ci-DME) in patients within a year. The first approach involves using ResNet101 with cross-entropy as the loss function. The second approach utilizes InceptionV3 with a class-balanced focal loss function. The final approach employs ResNet152 with a class-balanced focal loss function and additional data augmentation techniques. Each approach is evaluated on its ability to predict ci-DME onset under the constraints of the DIAMOND challenge using DeepDRid synthetic dataset.

2 Method

2.1 Model

We had the possibility of making only three submissions. For our first submission, we used ResNet-101 with two fully connected (linear) layers. For the second submission, we used InceptionV3 with three fully connected layers with a class balanced focal loss. For the final submission, we finetuned ResNet-152 with three fully connected layers with a class balanced focal loss on augmented dataset. All the model are pre-trained on ImageNet Dataset. All the related code is made available on Github [12].

2.2 Loss Function

In the first submission, we used the cross-entropy loss function, which is widely adopted for multi-class classification tasks in deep learning [13]. Cross-entropy quantifies the divergence between the predicted probability distribution and the true distribution. For a classification problem with n classes, the cross-entropy loss L is defined as:

$$L = -\sum_{i=1}^{n} y_i \log(p_i) \quad (1)$$

where y_i represents the true probability of class i (1 for the correct class and 0 otherwise), and p_i is the predicted probability of class i, typically obtained via a softmax activation.

This loss function penalizes significant deviations from the correct prediction, guiding the model to improve its accuracy during training. In this context, cross-entropy provided a baseline for comparison against more complex loss functions.

While for the second and the third submission, given the imbalanced nature of the data, we employ a loss function approach proposed in [14] known as Class-Balanced Loss. This loss function approach addresses the challenge of training on imbalanced datasets by introducing a weighting factor that is inversely proportional to the effective number of samples in each class. The effective number of samples refers to the expected volume of samples for a given class.

Let $L(p, y)$ denote a general loss function, where y is the ground-truth label and p represents the predicted class probabilities. Suppose the number of samples in class i is n_i. The effective number of samples for class i is defined as:

$$E_{n_i} = \frac{1 - \beta^{n_i}}{1 - \beta} \quad (2)$$

where:

$$\beta = \frac{N_i - 1}{N_i} \quad (3)$$

And N_i is a hyperparameter. As it is challenging to determine optimal hyperparameters for each class, the authors propose setting $N_i = N$ and $\beta = \frac{N-1}{N}$ uniformly across all classes in the dataset. Where N is the volume of all possible data in the feature space of a class.

Thus, the Class-Balanced (CB) loss can be expressed as:

$$CB(p, y) = \frac{1 - \beta}{1 - \beta^{n_y}} L(p, y) \quad (4)$$

where n_y is the number of samples in the ground-truth class y, and $\beta \in [0, 1)L$ is a hyper-parameter. This formulation enables the CB loss to smoothly adjust between no re-weighting (when $\beta = 0$) and full re-weighting by inverse class frequency (as β approaches 1).

Notably, this class-balanced term is independent of the choice of loss function L and the predicted class probabilities p. This makes it highly versatile and applicable to various deep learning tasks.

We chose the Focal Loss function, introduced in [8]. Focal Loss is specifically designed to focus the learning process on hard-to-classify examples, particularly those from minority classes, by down-weighting the contribution of easy examples.

The Focal Loss modifies the standard cross-entropy loss by introducing a modulating factor $(1 - p_t)^\gamma$, where p_t is the model's estimated probability for the true class and γ is a focusing parameter that adjusts the rate at which easy examples are down-weighted. The Focal Loss is defined as:

$$FL(z, y) = -\sum_{i=1}^{C}(1 - p_{t_i})^\gamma \log(p_{t_i}) \qquad (5)$$

where $\gamma \geq 0$ is a tunable focusing parameter. When $\gamma = 0$, Focal Loss reduces to the standard cross-entropy loss. As γ increases, the loss function increasingly focuses on hard-to-classify examples, effectively addressing the class imbalance problem by mitigating the influence of easily classified, majority class examples.

The Class-Balanced Focal Loss is given by:

$$CBFL(z, y) = \frac{1-\beta}{1-\beta^{n_y}}\sum_{i=1}^{C}(1 - p_{t_i})^\gamma \log(p_{t_i}) \qquad (6)$$

This combination ensures that the loss function remains robust against class imbalance while focusing on difficult-to-classify examples.

2.3 Data

In this challenge, participants are not given direct access to the training, validation, or test datasets. Instead, models must be submitted in the form of code, which is then executed by the organizers on a locally-based cluster. The training data used in this challenge is sourced from 14 French hospitals as part of the EviRed project [10], which aims to predict the development of ci-DME and anticipate the onset of DR complications. To evaluate the developed models, the DIAMOND Challenge also incorporates independent out-of-domain datasets from Algeria, in addition to the French hospital data.

For model development and local testing, the organizers provide two datasets: a synthetic dataset created from the Deep Diabetic Retinopathy Image Dataset (DeepDRiD) [11], and tow real images from a single patient from the EviRed dataset. The DeepDRiD dataset includes images from several diabetic complication screening projects in Shanghai, such as the Shanghai Diabetic Complication Screening Project, the Nicheng Diabetes Screening Project, and the Nationwide Screening for Complications of Diabetes. It comprises 2,000 regular fundus images from 500 patients (four images per patient, with two images per eye centered on the macula and optic disc) and 256 UWF retinal images from 128 patients (two images per patient, one per eye). Diagnostic labels for diabetic retinopathy are provided based on the International Clinical Diabetic Retinopathy (ICDR) classification, with grades ranging from no apparent retinopathy (Grade 0) to proliferative DR (Grade 4). The dataset is split into training, testing, and online validation sets for both regular fundus and UWF images. In the DeepDRiD synthetic dataset, the organizers only included UWF-CFP images, selecting 150 images for training and 50 for validation.

In addition to the DeepDRiD synthetic dataset, the organizers have provided four real images from a single patient (one OPTOS image and one CLARUS image for each eye). These images, shared under a CC BY-NC-ND license, are intended to enhance model development using actual ci-DME labels. They also aim to simulate the challenge's hidden dataset, which will be used for the final evaluation.

2.4 Evaluation Metrics

The evaluation of the models in this challenge will be based on three key metrics: the Area Under the Receiver Operating Characteristic Curve (AUC), the F1-score (F1), and the Expected Calibration Error (ECE). These metrics will be combined to determine the final ranking of the models using the following formula:

$$S = \text{AUC} + 0.5 \times \text{F1} + 0.5 \times (1 - \text{ECE}) \tag{7}$$

The Expected Calibration Error (ECE) measures the model's calibration, which refers to how well the predicted probabilities align with the true probabilities. In the context of the challenge, ECE assesses the model's ability to confidently and reliably predict the onset of CI-DME within a year. This metric ensures that the model's predictions are both accurate and trustworthy, meaning that the predicted probabilities reflect the true likelihood of the outcome. It is calculated by taking a weighted average of the absolute difference between accuracy (acc) and confidence (conf) across a set of bins. The weight for each bin is proportional to the number of samples it contains. The formula for ECE is:

$$\text{ECE} = \sum_{i=1}^{B} \frac{|B_i|}{n} \times |\text{acc}(B_i) - \text{conf}(B_i)| \tag{8}$$

where B is the number of bins, $|B_i|$ is the number of samples in bin i, n is the total number of samples, $\text{acc}(B_i)$ is the accuracy of bin i, and $\text{conf}(B_i)$ is the confidence of bin i.

The F1-score, another key metric, represents the harmonic mean of precision and recall, offering a balanced measure of a model's performance, particularly when dealing with imbalanced classes. In the context of the challenge, the F1-score evaluates the model's ability to address the issue of data imbalance effectively. This is especially important because there is no prior information about the data distribution in the training and test sets. Additionally, the F1-score indirectly reflects the generalizability of the model, as it emphasizes the need to perform well on both precision and recall, regardless of the dataset's class proportions.

The F1-score is calculated as follows:

$$F1 = 2 \times \frac{\text{Precision} \times \text{Recall}}{\text{Precision} + \text{Recall}} \tag{9}$$

where precision is the ratio of true positives to the sum of true and false positives, and recall is the ratio of true positives to the sum of true positives and false negatives.

Finally, the AUC is used to measure the model's ability to distinguish between different classes. AUC is calculated as the area under the Receiver Operating Characteristic (ROC) curve, which plots the true positive rate (TPR) against the false positive rate (FPR). The AUC provides a single scalar value that reflects the overall performance of the model across all classification thresholds. It indicates how well the model distinguishes between patients who will develop CI-DME within the next year and those who will not.

2.5 Experimental Setting

For the submission 1, we used ResNet101 with a cross-entropy focal loss function. Data augmentation was limited to a single technique, where synthetic data was included, resulting in a training set of 154 images and a validation set of 50 images, with a batch size of 4. In the submission 2, we employed InceptionV3 as the backbone model and used a class-balanced focal loss function, setting the value of γ to 2 and β to 0.9. The same data augmentation technique was applied, and the dataset size remained unchanged, with a batch size of 4. For hyperparameter selection, we used the default values because we are unaware of the distribution of the organizer's dataset and the class distribution. Additionally, we lack sufficient computational resources to fine-tune the hyperparameters.

For the submission 3, we used ResNet152 as the backbone model, combined with a class-balanced focal loss function with same hyper-parameters setting. This submission involved two data augmentation techniques: a combination of Random Horizontal Flip and Random Vertical Flip, followed by a second augmentation using Random Crop Size and Gaussian Blur. These augmentations increased the dataset size to 462 images for training and 150 images for validation, with a batch size of 40. Figure 1 illustrates the pipeline for the three approaches.

All submissions were trained on images of size 448×448 for 100 epochs. The training and deployment were conducted using the Singularity image provided by the organizers, which included all necessary libraries. The training was performed on Kaggle's free GPU, and the configurations were left as default as specified by the organizers.

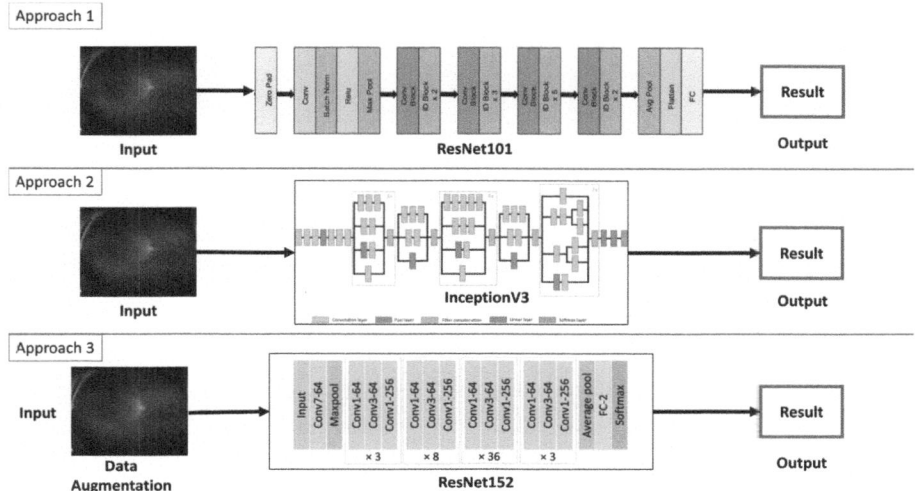

Fig. 1. The pipeline for the three approaches.

3 Results

Table 1 presents the results of the first and second submissions on EviRed dataset. The second submission, which involved using the InceptionV3 architecture and a class-balanced focal loss function, achieved the highest AUC of 0.53, with an F1 score of 0.14. However, it resulted in a poor ECE of 0.44, leading to an overall score of 0.88. In contrast, the first submission, which utilized ResNet101 with a cross-entropy loss function, produced the lowest F1 score of 0 but achieved a significantly better ECE of 0.07, leading to an overall score of 0.96.

Table 2 presents the performance results of the three approaches on the Deep-DRiD synthetic dataset. The third submission, which utilized ResNet152 with a class-balanced focal loss function and data augmentation, demonstrated the best overall performance, achieving an AUC of 0.79, the highest F1 score of 0.65, and the lowest ECE of 0.11. The second submission, while delivering a comparable F1 score of 0.57, exhibited the highest ECE at 0.17. In contrast, the first submission achieved the lowest F1 score of 0, with an ECE of 0.12. These results highlight the effectiveness of both model architecture and the use of data augmentation in improving predictive performance.

Table 1. The performance of our 3 approches after training for 100 epochs by the organiser on EviRed dataset.

	AUC	F1-Score	ECE	Score
submission 1	0.5	0.0	0.07	0.96
submission 2	0.53	0.14	0.44	0.88
submission 3	-	-	-	-

Table 2. The performance of our 3 approaches after training for 100 epochs on the DeepDRiD synthetic dataset.

	AUC	F1-Score	ECE	score
submission 1	0.73	0.0	0.12	1.17
submission 2	0.73	0.57	0.17	1.4
submission 3	0.79	0.65	0.11	1.56

4 Discussion

The results presented in the tables and performance graphs highlight the varying effectiveness of the three different submission approaches in this task.

For the submission 1, which used ResNet101 and the cross-entropy loss function, the model failed to perform effectively. The F1 score remained at 0 in both DeepDRiD synthetic dataset and the EviRed dataset. As observed in the Fig. 2, the curve for the EviRed was stable at 0 from the first step, while for the synthetic dataset, the F1 score fluctuated initially before stabilizing at 0. This suggests that the model encountered significant issues during training, likely due to the imbalanced data. Despite achieving a relatively low ECE of 0.12 on the synthetic data and 0.07 on the organizer's dataset, the model was unable to distinguish between classes, likely overfitting to the majority class.

The submission 2, using InceptionV3 and a class-balanced focal loss function, demonstrated improved performance, as evidenced by the Fig. 2 for the DeepDRiD synthetic dataset and the organizers' reports on the leaderboard. Although we lack figures for Submission 2 on the EviRed dataset, according to their leaderboard, the model performed better than Submission 1. For the DeepDRiD synthetic dataset, the F1 score fluctuated early during training but eventually stabilized at 0.57. On the EviRed dataset, the model stabilized at 0.14, which, while an improvement over the first submission, was still relatively low. This shows the positive impact of the class-balanced focal loss, which addresses imbalanced data issues, as well as the suitability of the InceptionV3 architecture. However, the higher ECE scores (0.17 on the DeepDRiD synthetic data and 0.4 on the EviRed dataset) suggest that the model was less well-calibrated, indicating some inconsistencies in learning.

The submission 3, employing ResNet152, the same class-balanced focal loss function, and data augmentation, achieved the highest overall performance on

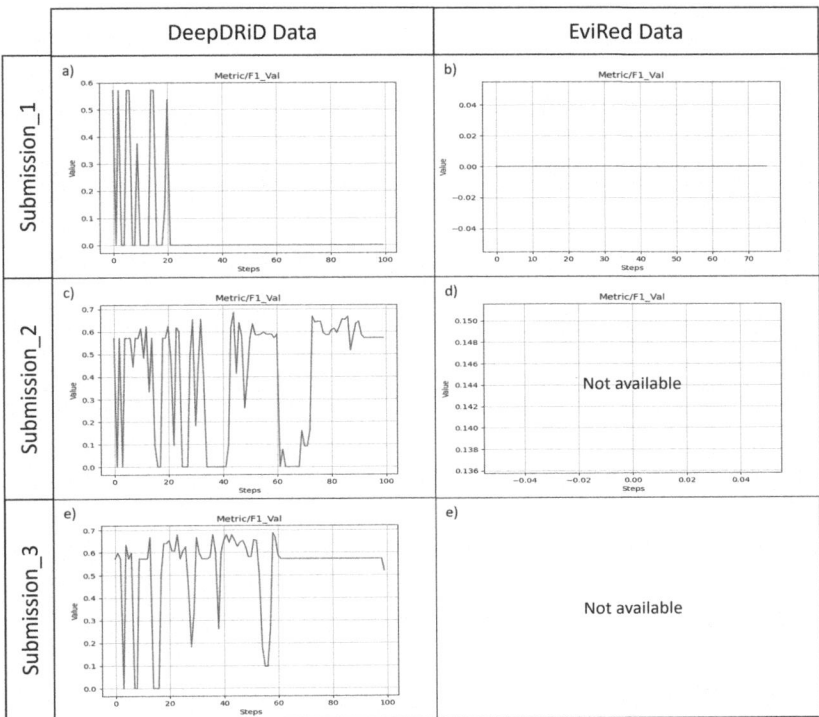

Fig. 2. The F1 score curves on the two validation datasets (Synthetic DeepDRiD and EVRID), where the left column represents the curves on the synthetic data and the right column shows the curves provided by the organizers on the EVRID dataset, over the epochs for all submissions. Figures (d) and (e) were not provided by the organizers.

the DeepDRiD synthetic dataset, applying data augmentation likely improved the model's ability to generalize and reduced the ECE compared to the submission 2. For the DeepDRiD synthetic data in Fig. 2 the last row, the F1 score started with fluctuations but stabilized significantly in the later stages, reaching a high of 0.65. Unfortunately, the results for the EviRed dataset were not reported on the leaderboard or logs. However, the performance on the synthetic data suggests that the combination of ResNet152, the class-balanced focal loss, and data augmentation resulted in more robust generalization. The high AUC score of 0.798 further supports this model's superior ability to discriminate between classes Fig. 3. Additionally, this submission achieved the lowest ECE 0.11 on the DeepDRiD synthetic dataset Fig. 4, indicating it was the best-calibrated model among the three submissions.

The comparative analysis of the graphs further reinforces these conclusions. The first submission, which showed low performance for the F1-score (as reflected in the first row of the Fig. 2), highlights the detrimental effects of limited augmen-

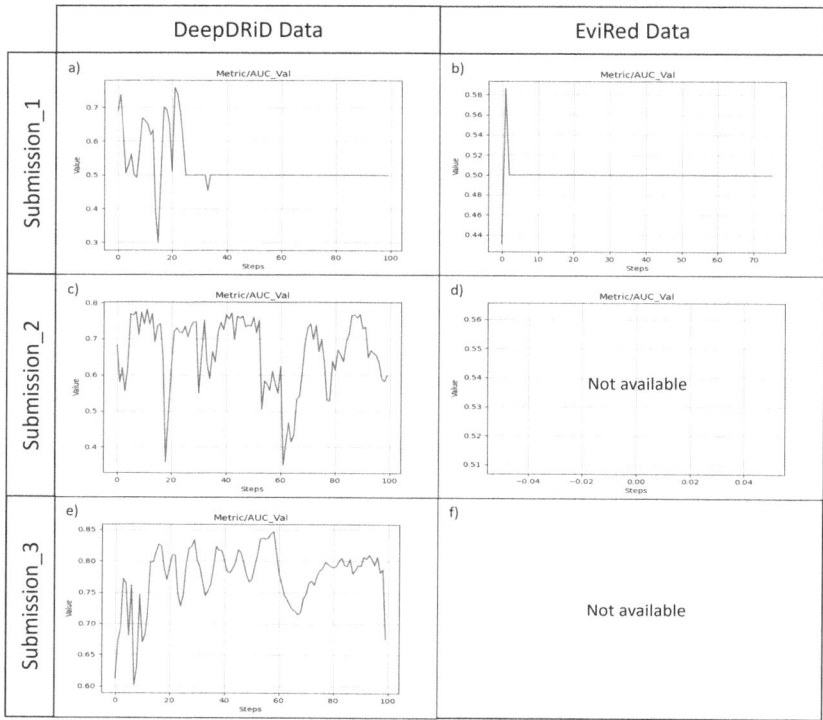

Fig. 3. The AUC curves on the two validation datasets (Synthetic DeepDRiD and EVRID), where the left column represents the curves on the synthetic data and the right column shows the curves provided by the organizers on the EVRID dataset, over the epochs for all submissions. Figures (d) and (e) were not provided by the organizers.

tation and an inadequate loss function, especially when working with imbalanced data.

The second submission, although demonstrating improved F1-score performance, exhibited some instability, as shown in Fig. 4, likely due to insufficient augmentation techniques.

In contrast, the second submission, with its stable F1-score on the DeepDRiD synthetic dataset, underscores the importance of incorporating effective data augmentation for model calibration, as supported by findings from another study [15]. Specifically, the augmentation techniques employed played a critical role in improving model performance.

The first augmentation technique included applying horizontal and vertical flips, which helped generate new examples and adjust the dataset's diversity. The second technique involved using random crop sizes followed by Gaussian blur, focusing on cropping the center of the image with varying sizes. This approach allowed the model to learn features essential for distinguishing between an eye

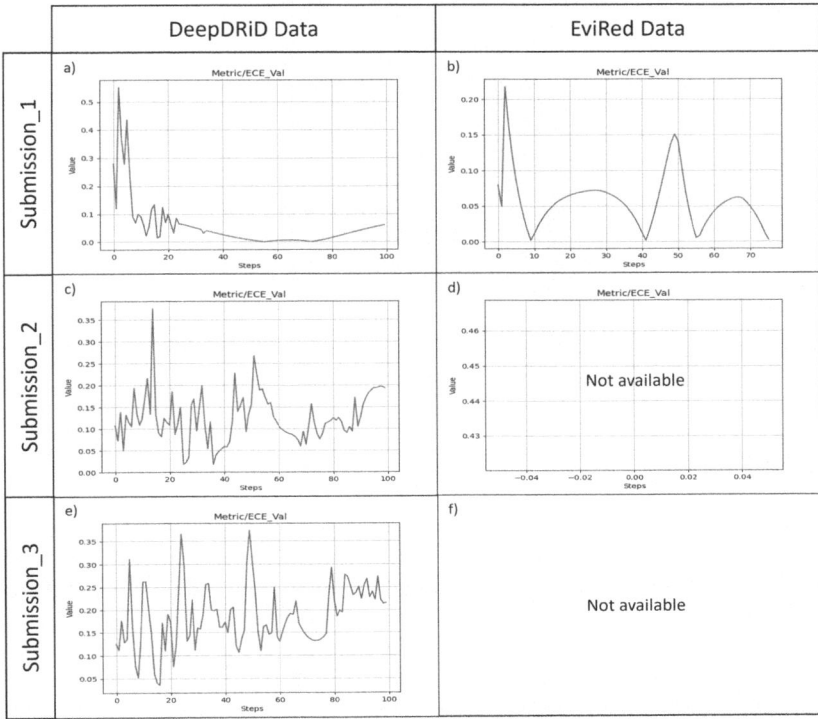

Fig. 4. The ECE score curves on the two validation datasets (Synthetic DeepDRiD and EVRID), where the left column represents the curves on the synthetic data and the right column shows the curves provided by the organizers on the EVRID dataset, over the epochs for all submissions. Figures (d) and (e) were not provided by the organizers.

at risk of developing CI-DME and a healthy eye, given that the disease often manifests in the central region of the image.

Additionally, the use of a balanced loss function contributed to developing a model that is not only stable and high-performing but also more generalizable.

Despite the overall performance of Submission 3, it reveals notable misclassifications, particularly in the left eye, where the model misclassified 44% of cases on the synthetic dataset. In contrast, the misclassification rate for the right eye was significantly lower at 19.2%.

The visual example Fig. 5 highlights that the UWF image of the patient's left eye shows no signs of CI-DME development within the next year, yet the model incorrectly predicted it as being at risk. This discrepancy suggests the presence of a bias in the model's predictions, emphasizing the need for further investigation and refinement to improve its robustness and accuracy.

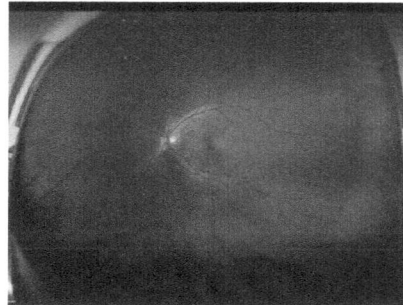

Fig. 5. Example of a misclassified UWF image: The model predicted this left eye as at risk of developing CI-DME within the next year, despite it is not.

5 Conclusion

In this work, we participated in a challenge organized to predict the occurrence of ci-DME within a year. The organizers provided a synthetic dataset derived from the DeepDRiD dataset for development purposes, while withholding the actual challenge dataset, which added an additional layer of complexity to the task. We submitted three models, each employing different architectures, loss functions, and data augmentation techniques.

The submission 1, using ResNet101 with a cross-entropy loss function, struggled significantly to generalize, as reflected by an F1 score of 0 across both the synthetic dataset and the organizer's dataset. Despite achieving low ECE scores, the model's poor performance can be attributed to its inability to handle the inherent data imbalance, likely leading to overfitting to the majority class.

The submission 2 improved on the first by utilizing InceptionV3 with a class-balanced focal loss function. This submission achieved a more favorable F1 score of 0.57 on the synthetic dataset and 0.14 on the EviRed dataset, demonstrating the potential benefits of focal loss in handling class imbalance. However, its relatively high ECE, particularly on the EviRed dataset (0.4), indicated suboptimal calibration and potential room for further improvement in model reliability.

The submission 3 and most successful submission incorporated ResNet152, class-balanced focal loss, and data augmentation. This model achieved the best overall performance with an F1 score of 0.65 on the DeepDRiD synthetic dataset, highlighting the positive impact of a more robust architecture and augmentation techniques. Additionally, this submission achieved the lowest ECE (0.11), indicating that it was the best-calibrated model, and an AUC score of 0.79, suggesting strong discriminatory power. Unfortunately, the performance on the EviRed dataset was not made available, but the DeepDRiD synthetic dataset results suggest that this model would perform well in real-world applications.

One of the challenges of the competition lies in the limitations of available information and the restricted access to the real dataset. As a result, the development of approaches must rely on the class distribution of the synthetic dataset. Despite these constraints, the competition encourages participants to design methods that can adapt and generalize effectively to other distributions, demonstrating robustness and versatility.

References

1. Wang, W., Lo, A.: Diabetic retinopathy: pathophysiology and treatments. Int. J. Mol. Sci. **19**(6), 1816 (2018)
2. Musat, O., et al.: Diabetic macular edema. Romanian J. Ophthalmol. **59**(3), 133–136 (2015)
3. Chouhan, S., et al.: Diagnostic utility of swept-source oct-based biometry and fundus photographs compared to spectral domain oct in center-involving diabetic macular edema. Ophthalmic Epidemiol. 1–8 (2024)
4. Miguel, A.R., Arruabarrena, C., Allendes, G., Olivera, M., Zarranz-Ventura, J., Teus, M.: Hybrid deep learning models for the screening of diabetic macular edema in optical coherence tomography volumes. Sci. Rep. **14** (2024)
5. Saidi, L., et al.: Automatic detection of AMD and DME retinal pathologies using deep learning. Int. J. Biomed. Imaging **2023**, 9966107 (2023)
6. Okonkwo, O.N., Hassan, A.O., Bogunjoko, T., Akinye, A., Akanbi, T., Agweye, C.: Low rates of optical coherence tomography utilization in the diagnosis and management of retinovascular diseases in a lower middle-income economy. Nigerian J. Clin. Pract. **26**(7), 1011–1016 (2023)
7. Varadarajan, A.V., et al.: Predicting optical coherence tomography-derived diabetic macular edema grades from fundus photographs using deep learning. Nat. Commun. **11**(1), 130 (2020)
8. Lin, T.-Y., Goyal, P., Girshick, R., He, K., Dollar, P.: Focal loss for dense object detection. IEEE Trans. Pattern Anal. Mach. Intell. (PAMI) (2018)
9. Device-independent diabetic macular edema onset prediction (diamond). Competition. Organized by Mostafa El Habib Daho (University of West Brittany, Brest, France & University of Tlemcen, Tlemcen, Algeria), Sarah Matta (University of West Brittany, Brest, France), Rachid Zeghlache (University of West Brittany, Brest, France), Alireza Rezaei (University of West Brittany, Brest, France), Mathias Gallardo (AP-HP, Paris, France and Fondation Rothschild, Paris, France), Capucine Lepicard (AP-HP, Paris, France), Mathieu Lamard (University of West Brittany, Brest, France), Pierre-Henri Conze (IMT Atlantique, Brest, France), Béatrice Cochener (University of West Brittany and Brest University Hospital, Brest, France), Aude Couturier (AP-HP, Paris, France), Sophie Bonnin (Fondation Rothschild, Paris, France), Ramin Tadayoni (AP-HP, Paris, France), and Gwenolé Quellec (Inserm, Brest, France)
10. Evired: Evaluation intelligente de la rétinopathie diabétique
11. Liu, R., et al.: Deepdrid: diabetic retinopathy-grading and image quality estimation challenge. Patterns **3**(6), 100512 (2022)
12. Lehmiani, Y., Elkhouzai, A., Mahmoudi, A.: MICCAI 2024 diamond challenge (2024). Accessed 20 Sept 2024
13. Mao, A., Mohri, M., Zhong, Y.: Cross-entropy loss functions: theoretical analysis and applications (2023)
14. Cui, Y., Jia, M., Lin, T.-Y., Song, Y., Belongie, S.: Class-balanced loss based on effective number of samples (2019)
15. Guo, C., Pleiss, G., Sun, Y., Weinberger, K.Q.: On calibration of modern neural networks (2017)

Calibrated Models for DME Progression Prediction from Ultra-Wide Field Retinal Images

Adrian Galdran[1,2]([✉])

[1] Computer Vision Center, Universitat Autònoma de Barcelona, Bellaterra, Spain
[2] Universitat Pompeu Fabra, Barcelona, Spain
adrian.galdran@upf.edu

Abstract. Diabetic Macular Edema (DME) is a leading cause of vision impairment in diabetic patients, requiring early diagnosis and progression monitoring. Optical Coherence Tomography (OCT) is often employed for detecting and monitoring DME, although standard retinal fundus images have been recently shown useful for this purpose too. Ultra-wide field (UWF) fundus imaging, an extension of conventional retinal imaging, offers a comprehensive view of the retina, providing a rich source of data for computational diagnostic models, potentially enabling the training of machine learning systems with high predictive performance. However, reliable predictions for clinical use require not only accurate models but also well-calibrated ones that provide trustworthy uncertainty estimates. This paper presents a solution to the DIAMOND challenge, a competition held in conjunction with MICCAI 2024, dealing with the task of DME progression forecasting from UWF images. Since one of the goals of DIAMOND is to assess model calibration, we explore various calibration techniques such as temperature scaling, deep ensembles, test-time augmentation or margin-based label smoothing, to improve predictive confidence estimates, a combination of which was used in our final submission. As DIAMOND did not release the training data (requiring code submission instead), we perform our analysis on the equally challenging task of glaucoma detection on a large-scale dataset, and report results for DIAMOND on a hidden validation set.

Keywords: DME Progression · Model Calibration · Ultra-Wide Field Retinal Imaging

1 Introduction

Diabetic macular edema (DME) is among the most relevant complications of diabetes [1]. This sight-threatening disease is typically encountered as thickening of the retina around the macula, often appearing with lesions like hard exudates. A computational tool capable of early diagnosis and progression forecasting of DME would represent a cost-effective solution for eye care in resource-deprived populations without access to ophthalmology specialists [24].

The presence of DME is typically detected by means of Optical Coherence Tomography (OCT) imaging. Recent studies have demonstrated a specific variant of the disease called center-involved DME can also be analyzed using conventional color fundus images (CFI) [21]. Ultra-Wide Field (UWF) CFIs, a modern form of CFIs with greater retinal scanning area (see Fig. 1), might be used to better capture early signs of DME, when compared to standard CFI, but this hypothesis is yet to be proved.

In the above context, the DIAMOND competition poses the challenge of developing machine learning models that can leverage UWF-CFIs for enhancing computational DME analysis. Additionally, DIAMOND does not propose to model the current status of the disease from a given UWF-CFI, but its future evolution, forecasting the appearance or not of ci-DME. A successful solution to this problem would have potential for improving early detection and treatment strategies, minimizing instances of vision loss.

A particularity of the DIAMOND competition is a focus on model calibration. This is an essential property of machine learning classifiers for medical image diagnosis, since it directly impacts the reliability and interpretability of predictions [4,9]. Well-calibrated models provide probability estimates that reflect the true likelihood of an event (*e.g.*, disease presence), enabling clinicians to make informed decisions based on the model's output. Poorly calibrated models, even if accurate, can either understate or overstate confidence, leading to misdiagnosis or inappropriate treatments. This can have serious consequences, including unnecessary interventions or missed diagnoses. In this paper, we concentrate ourselves on this aspect of the ci-DME forecasting problem, analyzing different calibration mechanisms, and ways for combining them.

A secondary methodological challenge in DIAMOND is the un-availability of training data. Instead or releasing labeled images for the participants to develop their predictive systems, the organization requested the submission of code that would be trained on a computing server in the cloud. While limiting in terms of model development (*e.g.* designing meaningful image pre-processing techniques is hardly feasible without access to training data), this scenario effectively simulates the difficulties of real-life clinical model development, arising from issues like data privacy or ethical considerations that might prevent sensitive data sharing.

2 Methodology

In this section, we first introduce some notation about classifiers and model calibration, then describe several popular calibration techniques, and give an overview of our approach to the DIAMOND challenge.

2.1 Definitions and Notation

Let us start with a binary classifier $\mathcal{M}$ trained on samples (x, y), where x is an image, and $y \in \{0, 1\}$ its associated category. Traditionally, binary classifiers are

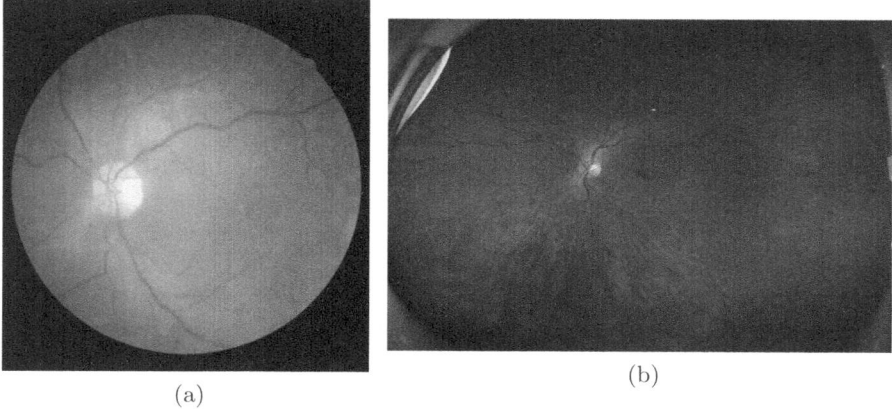

Fig. 1. A CFI (a) and corresponding UW-CFI (b) from DeepDRiD [13]), showing greater field of view, potentially leading to improved early detection of ci-DME.

designed as models with a single output as number, $\mathcal{M}(x) = s \in [0,1]$. This score is often interpreted as a probability for x to belong to the positive class, i.e. $s \approx p(y=1|x)$.

If model scores were truly aligned with probabilities, we could simply reach a binary decision by thresholding the output in $t = 0.5$. We could also use the score for downstream tasks, e.g. combining different models that solve other tasks; think, if the sample is deemed positive, we proceed to segment an object in the image, otherwise we discard it. Another benefit would be that model scores would be readily interpretable, and could be shown to end-users like clinicians who want to know the likelihood of a patient having a disease: it might well be that in some situations even a low-probability event should lead to taking action due to a particular risk assessment strategy.

Unfortunately, it is not always the case that classifier scores are aligned with true probabilities. Indeed, we can define miscalibration as the gap between model scores and actual probabilities. Miscalibration needs to be measured on held-out data that has not been used for training (where a model will be typically more accurate and confident, so errors will be biased). Starting from $X_{\text{test}} = \{(x_i, y_i)\}_{i=1}^N$, a simple way to measure calibration error it would be to just calculate the average difference between scores and probabilities, $\mathcal{E} = \frac{1}{N} \sum_i |s_i \approx p(y_i = 1|x_i)|$. Since we cannot estimate $p(y_i = 1|x_i)$ on a single item, we need to group samples before in bins. This gives rise to the definition of Expected Calibration Error, the most common measure of miscalibration:

$$\text{ECE} = \frac{1}{N} \sum_{i=1}^N \frac{1}{|B_i|} \|\bar{s}_i - \mathbb{1}_i\|, \qquad (1)$$

where B_i is a bin containing a subset of X_{test}, $\bar{s}_i$ is the average value of scores corresponding to samples in B_i, and $\mathbb{1}_i$ represent the proportion of samples in B_i belonging to the positive category.

A first observation is that, if scores truly captured likelihood of belonging to class 1, then we would have ECE = 0. Also note that, in order for a model to be calibrated, it does not need to be accurate. A classifier returning random predictions on a balanced test set would achieve zero calibration error, same as a perfect, always-correct classifier. In some sense, calibration and discrimination ability are orthogonal properties of classifiers.

2.2 Calibration Approaches and Their Combination

Once we understand how to measure the degree of calibration for a classifier, it makes sense to ask how can we increase it. Generally speaking, we can say that there are three kind of approaches to this problem: training-time calibration techniques, ensembling methods, and post-processing algorithms.

Training-Time Calibration. Modern neural networks can be expected to become miscalibrated, specifically overconfident in their predictions [5]. Since they are typically trained by minimizing a Negative Log-Likelihood loss on the training data, it is standard to keep pushing down the training loss, as extreme correct (or incorrect!) predictions will not be penalized by common performance metrics like accuracy, F1, or AUC [20]. Therefore, it becomes meaningful to intervene in the training dynamics in order to stop this phenomenon from happening.

A prominent family of methods is related to Label Smoothing [17]. This technique attempts to avoid over-confident predictions by artificially modifying the ground-truth values in such a way that a model will be penalized whenever it predicts a category with full-confidence. In the binary case, smoothing labels by an amount of 10% would amount to replacing $y = 0$ by $y = 0.1$ and $y = 1$ by $y = 0.9$. This technique has been extended in a number of different manners. For example, Label Smoothing has been applied to the multi-annotator scenario in [3], extended to segmentation problems [16], or modified as a category-dependent smoothing factor λ on training samples (Margin-Based Label Smoothing, [11,12]).

Other training-time methods involve manipulations of the loss function minimized during training, like explicit regularization terms to discourage a model from becoming overconfident [10], mix-up-like regularization strategies [25] or focal losses [15].

Model Ensembling. Standard Model Ensembling has been long known to bring improvements both in the discrimination and in the calibration sides of predictive models [8]. For obvious reasons, training an ensemble of N different models results in a process N times more computationally expensive than training a single model. Motivated by this, there has been intense research on "weak-ensembles", *i.e.* strategies to generate multiple plausible predictions than can be later averaged, without having to train multiple models from scratch. Among the most popular ones are Test-Time Dropout (carry out multiple forward passes in test time, without deactivating the drop-out layers of a model, [2]) or Test-Time Augmentation (making several forward passes, but applying data augmentation to the input image and then collecting the resulting predictions, [23]). Other similar techniques involve training a single model with multiple classifier heads [4], or extracting predictions from different layers of the network [7].

Calibration by Post-processing. The last type of calibration techniques deals with already-trained models. This has the advantage of being computationally cheap, and be considered most of the times as designing "model wrappers". The most popular technique in this family is Temperature Calibration [5], which attempts to dampen the pre-activation outputs of a deep neural network by means of a simple scalar multiplication. It important to remark that since the transform we apply is monotone, the rank of predictions will not be affected, *i.e.* ranking-based metrics like the AUC will remain unaltered after Temperature Scaling. Tempering outputs in this way was initially proposed for multiclass classifiers, and represents a relatively straightforward extension of an older technique designed for binary classifier known as Platt Scaling [19]. Another common post-processing algorithm is Isotonic Regression [18]. Lately, Conformal Prediction has also been suggested as a direction for improving model calibration after training [14].

2.3 The DIAMOND Challenge - Model Training and Inference Details

After some light model selection, we used a Swin transformer [6] as our binary classifier, trained in a five-fold manner. The models were optimized by minimizing the Cross-Entropy loss with Margin-Based Label Smoothing, using a standard Adam optimizer and a batch-size of 8 512×512 images, starting from a learning rate of $l = 2e-5$ that was decayed cyclically during 30 epochs, at intervals of 5 epochs. After each cycle, performance was monitored on a separate validation set in terms of AUC+AP (see next section for a discussion on metrics), and the best checkpoint was kept. Inference on images not used for training was conducted by applying Test-Time Augmentation (horizontal and vertical flips) to the images, and then generating predictions with the five available models. The resulting probabilities were all averaged to obtain a final probability for each class, see next paragraph for some comments on the assigned category.

A Note on Model Output. Up to now, we have assumed that a binary classifier would have a single number as output, $\mathcal{M}(x) = s \in [0, 1]$. However, the nature of the DIAMOND challenge forced participants to produce an output

$\mathcal{M}(x) = (s_0, s_1) \in [0,1] \times [0,1]$, with $\sum_i s_i = s_0 + s_1 = 1$, which is the typical setup of a multi-class classifier with $K > 2$ categories. The pre-activation outputs typically undergo a softmax operation to satisfy this constraint. Unfortunately, under this formulation the final decision of the classifier is automatically defined as the category with maximum allocated value, $i.e.$ $\hat{y} = \mathrm{argmax}(s_0, s_1)$. In contrast, for a single output, one can always fix a threshold different than $t = 0.5$ when producing the final prediction. This circumstance did not affect the competition metric on its AUC and ECE components, as they are threshold-independent quantities. However, it was detrimental for the F1 metric, as due to class imbalance, most examples were always assigned the negative category after applying an arg-max operation on the output of submitted models. Lowering the decision threshold would have had a beneficial impact in our final result.

3 Experimental Analysis

3.1 Datasets and Performance Evaluation

One of the particularities of the DIAMOND challenge was that there was no training data to use for model development. Instead, data remained on the organization side, and participants submitted code that could be ran by the organizers to generate validation metrics. Only a few code submissions against training and validation data were accepted by the organization, and a final one against test data. The only information available at the time of writing about the "hidden" dataset is that training data was originated from 14 French institutions, and that the test set would incorporate additional data sourced from Algeria. Images from a single patient were provided, in order to have a better understanding of the nature of the challenge data, see Fig. 2.

Fig. 2. UW-CFI images provided by the organizers, (a) and (b) show images acquired with an OPTOS and a CLARUS device respectively. It can be appreciated that the diversity in acquisition devices results in wildly different image appearances in the training set. Only imaging from one patient was provided to participants.

In order to guide model development, we decided to switch to a different diagnostic task, Glaucoma prediction from eye fundus images. For this, we used a subset of the AIROGS dataset [22], since it contains a great amount of diverse data, with a remarkable class imbalance. Performance was assessed by means of the Area Under the AUC curve (AUROC), and also the area under the Precision-Recall curve (Average Precision, AP), which is more meaningful for class-imbalanced scenarios[1]

3.2 Numerical Results and Discussion

For the AIROGS experiments, data was split in five subsets and each possible combination of four of them was used for training a single model. We also computed model ensembling performance, and the result of adding Temperature Scaling. In addition, we enable test-time data augmentation. On this dataset, we conducted a hyperparameter selection in order to find out the most appropriate value for the margin λ in the Margin-Based Label Smoothing scheme. The result of the these experiments is shown on Table 1. There are some interesting conclusions that we can draw from it:

Table 1. Cross-Validation results on the AIROGS dataset for glaucoma prediction for different margin values in Margin-Based Label Smoothing (**MbLS**), using also Model Ensembling (**Ensemble**) and Temperature Scaling (**TS**).

	MbLS $\lambda = 6$			MbLS $\lambda = 8$			MbLS $\lambda = 10$		
	AUC	AP	ECE	AUC	AP	ECE	AUC	AP	ECE
Fold 1	97.32	85.63	1.1169	97.58	86.40	1.6565	96.81	86.22	1.3300
Fold 2	97.12	86.26	1.5329	97.38	86.63	2.0872	97.23	85.51	1.7081
Fold 3	97.30	86.77	1.4338	96.93	85.50	0.9770	97.57	86.22	0.9820
Fold 4	97.13	86.34	1.1589	96.95	86.26	1.5459	97.64	86.41	1.3621
Fold 5	97.23	86.15	1.6713	96.91	86.00	1.7952	96.81	85.45	1.4822
Ensemble	97.78	88.51	1.4069	97.72	88.36	1.4882	97.76	88.13	1.2412
+ TS	97.78	88.51	1.2076	97.72	88.36	0.7648	97.76	88.13	0.8476

1. **Ensembling always brings improvements:** It is clear that model ensembling provides improvements in terms of AUC, AP, and ECE in almost all cases, for all values of λ.
2. **Ensembling cancels fold-wise performance variability for AUC and AP.** Even if fold performances have some variations, the ensembling results are quite similar in terms of AUC and AP between models trained with different values of λ.

---
[1] The challenge scored solutions with a combination of AUC, ECE and F1 metrics, but we prefer to use the AP instead of the F1 score for model selection purposes, since it is a threshold-less metric [20].

3. **Different margin values impact calibration error but not discrimination ability.** Although different values of λ do not appear to impact the AUC and AP scores of the models, their calibration error is indeed quite different: for $\lambda = 10$ we achieve a lower value ECE, 1.2412, while for $\lambda = 8$ this value increases up to ECE=1.4882.
4. **Temperature Scaling always brings improvements:** All models benefit from applying TS after ensembling. In addition, the worst calibrated model pre-TS becomes the best calibrated one afterwards.

We eventually selected the ensemble of models trained with $\lambda = 8$, since AUC and AP were similar to the other ones, but the ECE after TS was lower. We submitted code to recreate this five fold training on the "hidden dataset", although instead of splitting the training data we simply used a different random seed for each ensemble member (the train/validation split was fixed by the organization). The results for each model, together with the model ensemble and TS performance on the validation set, were returned to us by the organization, and are shown on Table 2. These numbers are not readily comparable with the ones in Table 1, they do show that TS had a great impact in reducing the calibration error. On the other hand, ensembling in this case did not appear to help in improving discrimination metrics like AUC and AP.

The same code submission was used to generate predictions on a final hidden test set, but metric values were not public at the time of writing.

Table 2. Validation results obtained in the DIAMOND dataset for a five-fold "blind" model training (code submission, no data available to participants), including Model Ensembling and Temperature Scaling.

	Fold 1	Fold 2	Fold 3	Fold 4	Fold 5	Ens.	Ens. + TS
AUC	70.10	70.52	67.85	71.40	72.04	69.49	69.49
AP	20.35	16.46	15.69	18.98	18.52	18.66	18.66
ECE	5.7957	2.2338	5.3774	5.0687	4.9210	4.6905	1.1857

4 Conclusions

In this paper, we focused on the calibration properties of machine learning models trained to predict Diabetic Macular Edema (DME) progression from ultra-wide field (UWF) fundus images. While deep learning models are capable of delivering high diagnostic accuracy, achieving model calibration significantly enhances their reliability and clinical utility. By applying techniques such as label smoothing, temperature scaling, or deep ensembles, we were able to improve the alignment between predicted probabilities and true outcomes on both a DME and a separate glaucoma detection task.

We developed our models in the context to the DIAMOND competition, which required code submission and did not share training data. Although this situation mimics some of the logistic difficulties associated with designing machine learning models in a realistic clinical setting, it did hinder model performance, specially due to the inability to implement tailored data augmentation strategies, perform error analysis, and design effective image pre-processing operations. Future work will deal with all this aspects of the problem, in order to maximize both model calibration and predictive performance.

Acknowledgments. A. Galdran is funded by a Ramon y Cajal fellowship RYC2022-037144-I.

References

1. Chen, X., et al.: Diabetes mellitus and risk of age-related macular degeneration: a systematic review and meta-analysis. PLoS ONE **9**(9), e108196 (2014). https://doi.org/10.1371/journal.pone.0108196
2. Gal, Y., Ghahramani, Z.: Dropout as a Bayesian approximation: representing model uncertainty in deep learning. In: Proceedings of The 33rd International Conference on Machine Learning, pp. 1050–1059. PMLR (2016). iSSN 1938-7228
3. Galdran, A., Ballester, M.A.G.: Data-centric label smoothing for explainable glaucoma screening from eye fundus images (2024). https://doi.org/10.48550/arXiv.2406.03903
4. Galdran, A., Verjans, J.W., Carneiro, G., González Ballester, M.A.: Multi-head multi-loss model calibration. In: Greenspan, H., et al. (eds.) MICCAI 2023. LNCS, vol. 14222, pp. 108–117. Springer, Cham (2023). https://doi.org/10.1007/978-3-031-43898-1_11
5. Guo, C., Pleiss, G., Sun, Y., Weinberger, K.Q.: On calibration of modern neural networks. In: Proceedings of the 34th International Conference on Machine Learning, pp. 1321–1330. PMLR (2017)
6. Hatamizadeh, A., Nath, V., Tang, Y., Yang, D., Roth, H.R., Xu, D.: Swin UNETR: swin transformers for semantic segmentation of brain tumors in MRI images. In: Crimi, A., Bakas, S. (eds.) BrainLes 2021. LNCS, vol. 12962, pp. 272–284. Springer, Cham (2022). https://doi.org/10.1007/978-3-031-08999-2_22
7. Kushibar, K., Campello, V., Garrucho, L., Linardos, A., Radeva, P., Lekadir, K.: Layer ensembles: a single-pass uncertainty estimation in deep learning for segmentation. In: Wang, L., Dou, Q., Fletcher, P.T., Speidel, S., Li, S. (eds.) MICCAI 2022. LNCS, vol. 13438, pp. 514–524. Springer, Cham (2022). https://doi.org/10.1007/978-3-031-16452-1_49
8. Lakshminarayanan, B., Pritzel, A., Blundell, C.: Simple and scalable predictive uncertainty estimation using deep ensembles. In: Guyon, I., Luxburg, U.V., Bengio, S., Wallach, H., Fergus, R., Vishwanathan, S., Garnett, R. (eds.) Advances in Neural Information Processing Systems, vol. 30. Curran Associates, Inc. (2017)
9. Lambert, B., Forbes, F., Doyle, S., Dehaene, H., Dojat, M.: Trustworthy clinical AI solutions: a unified review of uncertainty quantification in deep learning models for medical image analysis. Artif. Intell. Med. **150**, 102830 (2024). https://doi.org/10.1016/j.artmed.2024.102830

10. Larrazabal, A.J., Martínez, C., Dolz, J., Ferrante, E.: Maximum entropy on erroneous predictions: improving model calibration for medical image segmentation. In: Greenspan, H., et al. (eds.) MICCAI 2023. LNCS, vol. 14222, pp. 273–283. Springer, Cham (2023). https://doi.org/10.1007/978-3-031-43898-1_27
11. Liu, B., Ayed, I.B., Galdran, A., Dolz, J.: The devil is in the margin: margin-based label smoothing for network calibration. In: 2022 IEEE/CVF Conference on Computer Vision and Pattern Recognition (CVPR), pp. 80–88 (2022). https://doi.org/10.1109/CVPR52688.2022.00018
12. Liu, B., Rony, J., Galdran, A., Dolz, J., Ben Ayed, I.: Class adaptive network calibration. In: Proceedings of the IEEE/CVF Conference on Computer Vision and Pattern Recognition, pp. 16070–16079 (2023)
13. Liu, R., et al.: DeepDRiD: diabetic retinopathy-grading and image quality estimation challenge. Patterns **3**(6) (2022). https://doi.org/10.1016/j.patter.2022.100512
14. Marx, C., Zhao, S., Neiswanger, W., Ermon, S.: Modular conformal calibration. In: Chaudhuri, K., Jegelka, S., Song, L., Szepesvari, C., Niu, G., Sabato, S. (eds.) Proceedings of the 39th International Conference on Machine Learning. Proceedings of Machine Learning Research, vol. 162, pp. 15180–15195. PMLR (2022)
15. Mukhoti, J., Kulharia, V., Sanyal, A., Golodetz, S., Torr, P., Dokania, P.: Calibrating deep neural networks using focal loss. In: Advances in Neural Information Processing Systems, vol. 33, pp. 15288–15299. Curran Associates, Inc. (2020)
16. Murugesan, B., Liu, B., Galdran, A., Ayed, I.B., Dolz, J.: Calibrating segmentation networks with margin-based label smoothing. Med. Image Anal. **87**, 102826 (2023). https://doi.org/10.1016/j.media.2023.102826
17. Müller, R., Kornblith, S., Hinton, G.E.: When does label smoothing help? In: Advances in Neural Information Processing Systems, vol. 32 (2019)
18. Niculescu-Mizil, A., Caruana, R.: Predicting good probabilities with supervised learning. In: Proceedings of the 22nd International Conference on Machine Learning, pp. 625–632 (2005)
19. Platt, J.C.: Probabilistic outputs for support vector machines and comparisons to regularized likelihood methods. In: Advances in Large Margin Classifiers, pp. 61–74. MIT Press (1999)
20. Reinke, A., et al.: Understanding metric-related pitfalls in image analysis validation. Nat. Methods **21**(2), 182–194 (2024). https://doi.org/10.1038/s41592-023-02150-0
21. Varadarajan, A.V., et al.: Predicting optical coherence tomography-derived diabetic macular edema grades from fundus photographs using deep learning. Nat. Commun. **11**(1), 130 (2020). https://doi.org/10.1038/s41467-019-13922-8
22. de Vente, C., et al.: AIROGS: artificial intelligence for robust glaucoma screening challenge. IEEE Trans. Med. Imaging **43**(1), 542–557 (2024). https://doi.org/10.1109/TMI.2023.3313786
23. Wang, G., Li, W., Aertsen, M., Deprest, J., Ourselin, S., Vercauteren, T.: Aleatoric uncertainty estimation with test-time augmentation for medical image segmentation with convolutional neural networks. Neurocomputing **338**, 34–45 (2019). https://doi.org/10.1016/j.neucom.2019.01.103
24. Zang, P., Hormel, T.T., Hwang, T.S., Bailey, S.T., Huang, D., Jia, Y.: Deep-learning-aided diagnosis of diabetic retinopathy, age-related macular degeneration, and glaucoma based on structural and angiographic OCT. Ophthalmol. Sci. **3**(1), 100245 (2023). https://doi.org/10.1016/j.xops.2022.100245
25. Zhang, J., Kailkhura, B., Han, T.: Mix-n-match: ensemble and compositional methods for uncertainty calibration in deep learning. In: International Conference on Machine Learning (ICML) (2020)

Deep Learning Ensemble for Predicting Diabetic Macular Edema Onset Using Ultra-Wide Field Color Fundus Image

Pengyao Qin[1](✉), Arun Thirunavukarasu[2], Theodoros Arvanitis[1], and Le Zhang[1]

[1] Digital Healthcare and Medical Imaging Research Group, School of Engineering, College of Engineering and Physical Sciences, University of Birmingham, Birmingham, UK
rqin135@qq.com
[2] Nuffield Department of Clinical Neurosciences, University of Oxford, Oxford, UK

Abstract. Diabetic macular edema (DME) is a severe complication of diabetes, characterized by thickening of the central portion of the retina due to accumulation of fluid. DME is a significant and common cause of visual impairment in diabetic patients. Center-involved DME (ci-DME) is the highest risk form of disease because fluid extends close to the fovea which is responsible for sharp central vision. Earlier diagnosis or prediction of ci-DME may improve treatment outcomes. Here, we propose an ensemble method to predict ci-DME onset within a year, after using synthetic ultra-wide field color fundus photography (UWF-CFP) images provided by the DIAMOND Challenge during development. We adopted a variety of baseline state-of-the-art classification networks including ResNet, DenseNet, EfficientNet, and VGG with the aim of enhancing model robustness. The best performing models were Densenet-121, Resnet-152 and EfficientNet-b7, and these were assembled into a definitive predictive model. The final ensemble model demonstrates a strong performance with an Area Under Curve (AUC) of 0.7017, an F1 score of 0.6512, and an Expected Calibration Error (ECE) of 0.2057 when deployed on the synthetic test dataset. Results from our ensemble model were superior/comparable to previous recorded results in highly curated settings using conventional fundus photography/ultra-wide field fundus photography. Optimal sensitivity in previous studies (using humans or computers to diagnose) ranges from 67.3%–98%, specificity from 47.8%–80%. Therefore, our method can be used safely and effectively in a range of settings may facilitate earlier diagnosis, better treatment decisions, and improved prognostication in ci-DME.

Keywords: Center-involved diabetic macular edema · Ultra-wide-field color fundus photography · Model ensemble · Deep learning

1 Introduction

Diabetic macular edema (DME) is a leading cause of vision loss among diabetic patients worldwide [1]. It is characterized by the accumulation of fluid in the central portion of the retina (the macula), and can occur at any stage in the natural history of diabetic retinopathy (DR) [2]. DME leads to severe vision impairment, although a variety of interventions are now available to prevent and even reverse disease progression if detected and treated early [2]. Center-involved diabetic macular edema (ci-DME) is the highest-risk form of disease, as the central macula (the fovea) is responsible for central sharp vision. Clinically, DME is best characterised using optical coherene tomography (OCT), as high resolution cross-sectional images of the retina clearly reveal fluid accumulation and can be measured to assess response to treatment.

En-face fundus photography, 2D colour imaging of the retina, is a cheaper and more widely available modality than OCT, and is already used in screening programmes to detect diabetic retinopathy [2,3]. However, clinicians cannot reliably diagnose DME on fundus photographs, and instead rely on surrogate markers of fluid accumulation such as hard exudate formation [4]. These surrogates are poor predictors of DME, and screening could therefore miss patients that would benefit from early treatment, while simultaneously resulting in many unnecessary referrals to ophthalmologists for further investigation [5]. Computational analysis of fundus photographs can improve on clinical assessment, and thereby improve the utility of fundus photography to screen for DME.

Ultra-wide field color fundus photography (UWF-CFP) is a more advanced form of en-face fundus imaging that captures a wider portion of the retina. UWF-CFP may capture more features of DME than conventional photography, such as through higher resolution imaging of the macula or due to peripheral retinal consequences of mass effect exerted by fluid accumulation. The DIAMOND Challenge aimed to leverage this potential through development of artificial intelligence models capable of predicting the onset of ci-DME within one year based on individual UWF-CFP images. By focusing on predictive modeling, the challenge sought to shift the management paradigm from reactive to proactive treatment of ci-DME, thereby reducing the incidence of vision loss in diabetic retinopathy [6]. The DIAMOND Challenge also introduced an additional methodological complexity: only code (rather than model weights) was submitted, which the organizing committee then ran on a cloud-based cluster. This tasked participants with developing highly generalizable models with the necessary flexibility to perform in real-world situations where data heterogeneity, privacy, and logistical limitations are common.

In this work, we propose a deep learning ensemble-based approach to predict the development of clinically significant diabetic macular edema (ci-DME) using ultra-widefield color fundus photography (UWF-CFP) images. To achieve this, we employed several state-of-the-art convolutional neural networks (CNNs), including DenseNet, ResNet, EfficientNet, and VGG architectures, each of which offers unique strengths in feature extraction and classification. These models were trained with comprehensive data augmentation strategies to enhance robustness and reduce overfitting, particularly given the limitations of the dataset size and variability.

To further improve predictive performance, an ensemble method was introduced to combine predictions from the top-performing models. By leveraging the complementary strengths of individual networks, the ensemble approach aims to boost overall accuracy and reliability, offering a more generalizable solution for ci-DME prediction. This methodology addresses both classification and calibration challenges, ensuring that the predictions are not only accurate but also reliable for potential clinical applications. The ensemble strategy underscores the importance of combining diverse model architectures to achieve a more robust and clinically meaningful outcome.

2 Method

2.1 Dataset

The DIAMOND Challenge provided data sourced from 14 French hospitals as part of the EVIRED project, which aims to predict ci-DME development and anticipate the onset of DR complications. For evaluation, the DIAMOND Challenge incorporated independent datasets from Algeria alongside the French hospital data. This approach promoted universally applicable solutions, capable of serving diverse population groups and settings. During the coding period, a synthetic dataset generated using images from the Deep Diabetic Retinopathy Image Dataset (DeepDRiD) was provided to assist in the development and local testing of the algorithms. DeepDRiD is available under the Creative Commons Attribution Share Alike 4.0 International license, and served as a development and testing resource with ci-DME labels generated based on DR severity. This dataset includes 204 UWF-CFP in total, with 154 images partitioned for training and 50 images partitioned for validation. Also a one-patient dataset is provided, including real image data from one patient, with images taken by different UWF-CFP device (OPTOS and CLARUS) from each eye to enhance model development with actual ci-DME labels.

2.2 Baseline Network

We adopted four CNN architectures to develop a generalisable and robust model: Resnet, Densenet, EfficientNet and VGG.

ResNets (Residual Networks) are renowned for high performance in classification tasks, particularly in medical image analysis. A deep residual learning framework is employed to address the vanishing gradient problem and the degradation of network performance as depth increases. This approach enables the training of very deep networks by introducing residual mapping, which allows gradients to propagate through the network more effectively, enhancing training efficiency and performance [7].

DenseNet (Densely Connected Convolutional Networks) excels in efficient feature usage and parameter reduction. It introduces dense blocks where each layer connects to every other layer in a feed-forward manner, promoting feature reuse and mitigating the vanishing gradient problem. DenseNet is often preferred for its ability to achieve high performance with fewer parameters than with ResNets [8].

EfficientNet scales network dimensions (depth, width, and resolution) uniformly using a compound scaling method. This model achieves superior performance on the ImageNet dataset and transfers well to other datasets, offering a balance of high accuracy and computational efficiency. EfficientNet's ability to maintain small model sizes while ensuring fast computation speeds makes it suitable for local applications such as hospital-based retinal image classification [9].

VGG (Visual Geometry Group Network) is an earlier convolutional neural network (CNN) architecture which uses small convolutional filters and deep networks (16–19 layers) to attain strong performance on large-scale image recognition tasks like ImageNet. VGG is known for its large model size and lengthy computation times relative to newer architectures, but remains a useful baseline model for image classification tasks such as object recognition and medical image classification [10].

2.3 Model Ensemble

Model ensembling combines the outputs produced by multiple models into a single prediction process, which can overcome shortcomings associated with individual estimators such as high variance, noise, and bias. In this work, three ensemble strategies were used.

Plurality Voting: The first was plurality voting: where the class with the highest number of votes is used as the final prediction. All models contribute equally to the decision-making process, preventing dominance by any single model. Plurality voting treats all votes equally, ignoring the confidence levels of individual model [11].

Averaging: The second was averaging: where the final outputted probability is the unweighted average of the probabilities estimated by each model. By averaging probabilities instead of class labels, this method incorporates the confidence of each model's predictions, often leading to better-calibrated outputs. Averaging can produce smoother predictions, especially in cases of highly imbalanced data, as it reduces the influence of extreme or outlier predictions from individual models [12].

Label Fusion: The third technique was label fusion using a three-layer simple neural network, which was trained using the predictions from several individual models as inputs, thereby learning to assign appropriate weights to each individual model. An example of these three strategies using five models and two classes is shown in Fig. 1. Final prediction outputs can vary widely with different ensemble strategies.

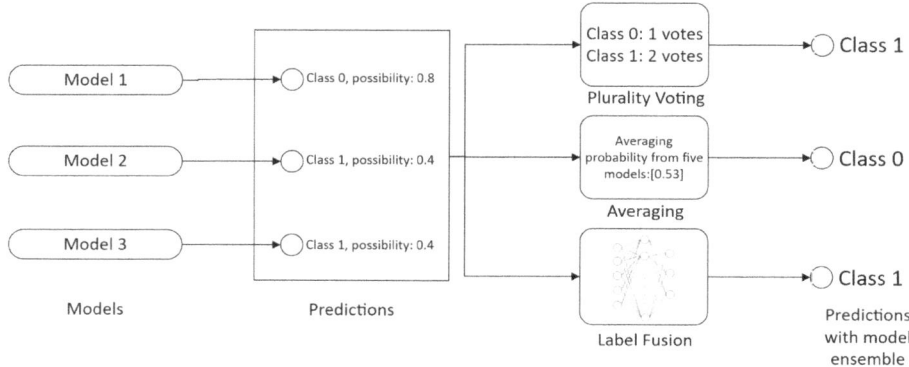

Fig. 1. Diagram of an example of each ensemble strategy used in this study, with three individual models predicting ci-DME.

3 Experiments

3.1 Experimental Setting

Image Preprocessing. In all experiments, the images were resized to 224 × 224 pixels from their original size. Other image augmentation techniques were also employed, including random horizontal reflection, random vertical reflection, and random rotation. Example images are shown in Fig. 2.

Training Hyper-parameters. Throughout the training process, the adam optimizer was used with a learning rate scheduler with exponential decay. Networks were trained using 200 epochs and the models with the best performance were saved. Performance was gauged with the metrics suggested by the DIAMOND Challenge: Area under the receiver operating characteristic curve (AUC), F1-score (F1) and Expected Calibration Error (ECE) [13].

3.2 Evaluation Metrics

For evaluation, we used the following metrics provided by the challenge organizer:

Area Under the Curve (AUC): The AUC is computed as the area under the receiver operating Characteristic (ROC) curve. AUC values range between 0 to 1; with 1 representing perfect classification performance across all model thresholds. The use of AUC aligns with the goal of achieving high sensitivity and specificity in predictions and was therefore weighted highest during evaluation and ranking.

F1 Score and Expected Calibration Error (ECE): The F1 score and ECE were considered as secondary metrics (lower weighted) to further assess the relevance of computed probabilities. The F1 score is the harmonic mean of precision and recall, which is used to balance these two indicators. F1 helped ensures that

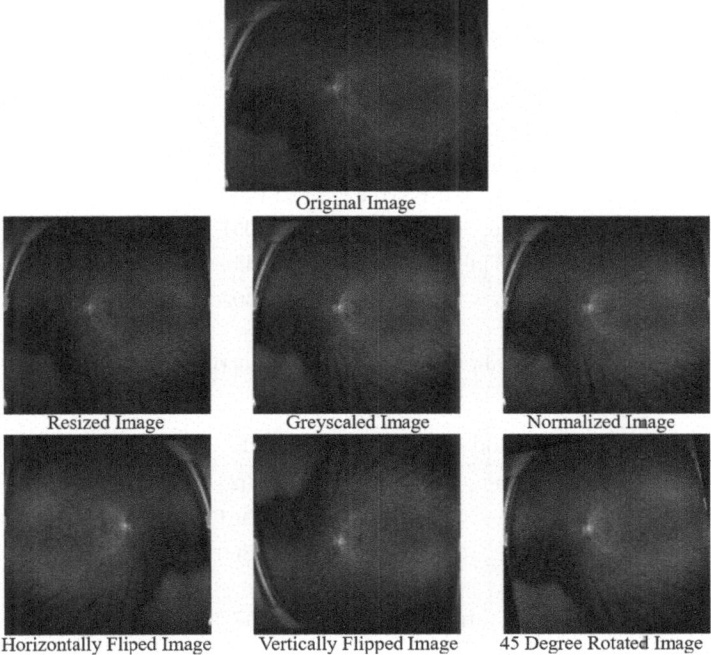

Fig. 2. Examples of image processing employed during model development. Images were resized to 224 × 224 px, greyscaled, normalized, randomly reflected in the horizontal and vertical axis, and randomly rotated through 45°. The effects of each step are shown in each panel.

binary predictions (when using a probability cutoff of 0.5) were relevant. F1 also assesses precision (or positive predictive value) which is not captured by AUC [14]. ECE measures the calibration of predicted probabilities, indicating the difference between predicted confidence and actual correctness. The calculation method of ECE is to divide the prediction distribution of the model output into multiple intervals, calculate the difference between the average prediction probability and the actual accuracy in each interval, and then sum them up to get the ECE value, lower ECE values indicate better calibration of the model. The use of ECE helped calibrate model properties for the context in which it will be used, such as by ensuring that its predicted probabilities matched the actual probabilities of the ground truth distribution [13]. The following performance function was used to determine which model was saved.

$$S = AUC + 0.5 \cdot F1 + 0.5 \cdot (1 - ECE)$$

3.3 Model Evaluation

The results in Table 1 compare the performance of 5 baseline models, with the DenseNet-121 achieving the highest overall score (1.5262) due to its strong AUC

Table 1. Experimental results for each baseline model, with inidividual evaluation metrics presented alongside overall performance scores. Densenet-121 exhibited the highest performance score.

Network	AUC	F1 Score	ECE	Overall Score
Resnet-50 [7]	0.7233	0.7778	0.3346	1.4449
Resnet-152 [7]	0.7633	0.6275	0.1966	1.47875
Densenet-121 [8]	0.7617	0.7347	0.2057	1.5262
EfficientNet-b7 [9]	0.7467	0.6780	0.2449	1.46325
VGG-19 [10]	0.7350	0.6154	0.3392	1.4231

Table 2. Experiment results about the evaluation metrics and overall scores of each model ensemble methods.

Ensemble method	AUC	F1 Score	ECE	Overall Score
Plurality voting	0.6517	0.5714	0.1023	1.3862
Averaging	0.71885	0.6862	0.5361	1.2939
Label Fusion	0.7017	0.6512	0.2860	1.4343

(0.7617), F1 score (0.7347), and low ECE (0.2057), reflecting its efficiency in classification and calibration. ResNet-152 also performed well, with the highest AUC (0.7633) and a solid overall score (1.47875), although its lower F1 score (0.6275) limited its effectiveness. EfficientNet-b7 demonstrated a balance between metrics, with a respectable overall score of 1.46325, but was outperformed by DenseNet-121. ResNet-50 and VGG-19, while adequate, lagged behind in overall scores (1.4449 and 1.4231, respectively), showing limitations in calibration and predictive accuracy compared to the more modern architectures.

Table 2 shows the performances of the ensemble methods, evaluated on the internal testing dataset. For the individual evaluation, label fusion obtained the best performances (overall score of 1.4343) comparing to the other models ensemble methods. The results in Table. 2 highlight the strengths and weaknesses of three ensemble methods for ci-DME prediction. Plurality voting, while simple and well-calibrated (ECE = 0.1023), achieved the lowest AUC (0.6517), F1 score (0.5714), and overall score (1.3862), indicating limited predictive capability. Averaging improved AUC (0.71885) and F1 score (0.6862) significantly, showcasing better classification performance, but it suffered from poor calibration (ECE = 0.5361), which lowered its overall score to 1.2939. Label fusion outperformed the other methods with the highest overall score (1.4343), offering a balanced performance with competitive AUC (0.7017), F1 score (0.6512), and improved calibration (ECE = 0.2860), demonstrating its ability to effectively integrate model outputs.

4 Discussion and Conclusion

In this study, we developed an ensemble-based deep learning model to predict the onset of ci-DME within 12 months using UWF-CFP images. Our model

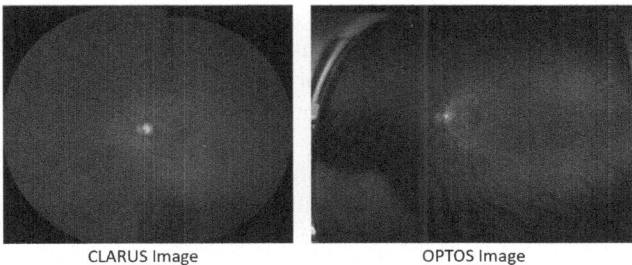

CLARUS Image   OPTOS Image

Fig. 3. Examples of Ultra Wide Field Colour Fundus Photographs captured by different devices, Zeiss CLARUS (left), and OPTOS (right). The Challenge uses datasets from these two devices for evaluation. Significant differences in dimensions, resolution, and artifact can confound classification and thereby place higher demands on the generalisability of a model that is planned to be trained and/or tested across both modalities.

combined several CNN architectures and exhibited performance comparable to previous studies involving human experts and individual deep learning models diagnosing existing DME on fundus photographs [1,5]. Our ensemble method ensured that classification decisions were not limited by any single model's weaknesses. This helped reduce the incidence of false negative classification, which in the real world could lead to patients missing out on crucial early treatment; as well as false positive classification, which can lead to overmedicalisation and unnecessary treatment. While the ensemble model's overall performance was slightly lower than that of the best individual model (Densenet-121), it exhibited improved generalizability across diverse datasets. This trade-off–sacrificing peak performance for better generalization–makes the ensemble model a more robust application for clinical deployment. Though higher performance has been achieved for detecting existing ci-DME from fundus photographs (rather than OCT, the clinical standard for diagnosis) [3], a predictive model like ours could help guide preemptive decision-making regarding active surveillance and early intervention. In the future, emerging attention-based models such as vision transformers (ViTs) could be integrated into the ensemble [15]. These models have demonstrated superior performance in image classification tasks driven by selective focus on relevant portions of the image, which is highly relevant to medical tasks like ci-DME prediction where pathological changes are localised.

In the DIAMOND Challenge, synthetic datasets were used exclusively for the initial training phase in which baseline networks were established. While this approach offers certain advantages, such as controlled data characteristics and reduced ethical concerns (particularly regarding data privacy), it introduces critical limitations that may limit model performance on real-world data. During model ensembling, certain strategies demonstrated poor results, which we hypothesize may be attributable to the characteristics of the synthetic data. For example: synthetic datasets may overemphasize specific features or lack challenging edge cases, leading to models that do not complement each other well in an ensemble. The absence of realistic noise in synthetic datasets can cause ensemble methods that rely on diverse model outputs to underperform, as the models may converge on similar, biased predictions [16]. To mitigate these limitations, future iterations of the study should incorporate real-world datasets in the training and validation stages.

Timely intervention is associated with substantial improvement in patient outcomes, but fundus photography is insufficient to replace OCT in diagnosing ci-DME and regular OCT screening is associated with prohibitive costs and inaccessibility, particularly in lower income countries [2,3]. Predictive deep learning models with UWF-CFP images may represent a cost-effective and scalable alternative, allowing more patients around the world to receive critical treatment as required [17]. This potential aligns well with the aims of the DIAMOND Challenge, which emphasized the value in generalisable models with potential to augment real-world clinical practice. Our approach aimed to maintain sufficient flexibility to cope with training and deployment across conventional fundus photography as well as UWF-CFP, with differences in frame of view, image artifact, and resolution that can confound classification (as shown in Fig. 3). In the 7-day period provided by the DIAMOND Challenge, we implemented a streamlined training and evaluation workflow using a bash script. This script automated the training of each component network separately on the available data. By using an ensemble method combined with model selection based on objective evaluation metrics, we ensured that the final submission was both robust and generalizable, potentially performing well even when applied to datasets derived from other imaging devices.

In conclusion, our ensemble model demonstrates the potential to significantly advance the early detection of ci-DME through the use of synthetic data before local deployment with authentic data from patients. Improved predictive models could facilitate proactive management and prompt treatment to improve outcomes in patients with or at risk of ci-DME. By integrating deep learning models with UWF-CFP imaging, accurate and accessible screening could promote timely intervention and ultimately better clinical outcomes for patients with DR.

References

1. Manikandan, S., Raman, R., Rajalakshmi, R., Tamilselvi, S., Surya, R.: Deep learning-based detection of diabetic macular edema using optical coherence tomography and fundus images: a meta-analysis. Indian J. Ophthalmol. **71**, 1783–1796 (2023)
2. Cheung, N., Mitchell, P., Wong, T.Y.: Diabetic retinopathy. Nat. Rev. Dis. Primers. **2**, 16012 (2016)
3. Varadarajan, A.V., et al.: Predicting optical coherence tomography-derived diabetic macular edema grades from fundus photographs using deep learning. Nat. Commun. **11**(1), 130 (2020)
4. Heng, L.Z., Collins, C., Ashraf, M., Chave, S., Scanlon, P.: Sensitivity of 2 dimensional color fundus photography surrogate markers as for diabetic macular Oedema. Invest. Ophthalmol. Visual Sci. **58**(8), 2914 (2017)

5. Wang, Y.T., Tadarati, M., Wolfson, Y., Bressler, S.B., Bressler, N.M.: Comparison of prevalence of diabetic macular edema based on monocular fundus photography vs optical coherence tomography. JAMA Ophthalmol. **134**(2), 222–228 (2016)
6. Gurung, R.L., et al.: Predictive factors for treatment outcomes with intravitreal anti-vascular endothelial growth factor injections in diabetic macular edema in clinical practice. Int. J. Retina Vitreous **9**(1), 23 (2023)
7. He, K., Zhang, X., Ren, S., Sun, J.: Deep residual learning for image recognition. In: Proceedings of the IEEE Conference on Computer Vision and Pattern Recognition, pp. 770–778 (2016)
8. Huang, G., Liu, Z., Van Der Maaten, L., Weinberger, K.Q.: Densely connected convolutional networks. In: Proceedings of the IEEE conference on computer vision and pattern recognition, pp. 4700–4708 (2017)
9. Tan, M., Le, Q.V.: EfficientNet: rethinking model scaling for convolutional neural networks. In: International Conference on Machine Learning, pp. 6105–6114. PMLR (2019)
10. Simonyan, K., Zisserman, A.: Very deep convolutional networks for large-scale image recognition. In: International Conference on Learning Representations, pp. 1–14 (2015)
11. Jain, A., Kumar, A., Susan, S.: Evaluating deep neural network ensembles by majority voting cum meta-learning scheme. In: Sivakumar Reddy, V., Kamakshi Prasad, V., Wang, J., Reddy, K.T.V. (eds.) Soft Computing and Signal Processing, pp. 29–37. Springer, Singapore (2022)
12. Ganaie, M.A., Hu, M., Malik, A.K., Tanveer, M., Suganthan, P.N.: Ensemble deep learning: a review. Eng. Appl. Artif. Intell. **115**, 105151 (2022)
13. Posocco, N., Bonnefoy, A.: Estimating expected calibration errors. In: Farkaš, I., Masulli, P., Otte, S., Wermter, S. (eds.) Artificial Neural Networks and Machine Learning – ICANN 2021, pp. 139–150. Springer, Cham (2021)
14. Grandini, M., Bagli, E., Visani, G.: Metrics for multi-class classification: an overview. ArXiv, abs/2008.05756 (2020)
15. Khan, A., et al.: A survey of the vision transformers and their CNN-transformer based variants. Artif. Intell. Rev. **56**, 1–54 (2023)
16. Hao, S., et al.: Synthetic data in AI: challenges, applications, and ethical implications. ArXiv, abs/2401.01629 (2024)
17. Tan, T., et al.: Artificial intelligence and digital health in global eye health: opportunities and challenges. Lancet Glob. Health **11**, E1432–E1443 (2023)

MARIO Challenge

Deep Learning Approaches for Monitoring Age-Related Macular Degeneration Progression in Optical Coherence Tomography

Yiding Hao(✉)

Imperial College London, London, UK
scyyh123@gmail.com

Abstract. Age-related Macular Degeneration (AMD) is a leading cause of visual impairment and severe vision loss, with an estimated 288 million patients expected by 2040. In recent years, significant advancements in Optical Coherence Tomography (OCT) technology have been made, increasing the professional demands on medical practitioners. Concurrently, the development of deep learning has shown great potential as a powerful tool in diagnosing such diseases. Under this background, the Monitoring Age-related Macular Degeneration Progression In Optical Coherence Tomography (MARIO) project was proposed, which includes two tasks: classify evolution between two pairs of 2-D slices from two consecutive 2D OCT acquisitions and prediction of evolution within 3 months of AMD on OCT 2D slices. This research work proposes two deep learning algorithms to accomplish both tasks. For Task 1, we implemented a Siamese network, and for Task 2, we employed a standalone network architecture combined with a post-processing method. Both network backbones are based on ConvNeXt-large and incorporate a multi-head attention mechanism. Our algorithms demonstrated good performance on the validation set, with an average score of 0.80427 for Task 1 and 0.43726 for Task 2, ranking third during the development phase. The code is available at https://github.com/lumine-1/MARIO_Project.

Keywords: Deep learning · Age-related macular degeneration · Optical coherence tomography · Medical image analysis

1 Introduction

Age-related macular degeneration is a leading cause of visual impairment and significant vision loss [1]. As of 2017, the global prevalence of AMD was 8.7%, and it is projected to reach 288 million by 2040 [2]. AMD is manifested by the accumulation of extracellular deposits, accompanied by progressive degeneration of adjacent tissues and photoreceptors [3]. Optical coherence tomography is a non-invasive technique that provides high-resolution cross-sectional images of the neurosensory retina and deeper structures [4]. It can also be used to measure the thickness of retinal cell layers as well as the entire retina [5]. For the diagnosis of AMD, OCT can display key diagnostic indicators such as structural changes in the retina and the separation of the Retinal Pigment Epithelium (RPE) from Bruch's membrane. Additionally, OCT can detect the accumulation of fluid between the retinal layers, aiding in the diagnosis of the type and severity of AMD [6]. Although OCT technology has made significant advancements in recent years, it has made interpretation of OCT images more complex, placing higher demands on medical practitioners [4].

Digital imaging and objective metrics are widely used in ophthalmology, making it well-suited for AI-assisted diagnosis. The rapid advancements in machine learning, particularly deep learning, have provided new methods for the automatic diagnosis of AMD [7]. For example, Convolutional Neural Networks (CNNs) have been frequently applied as a powerful tool for medical image analysis. AlexNet demonstrated the powerful capabilities of deep CNNs, attracting broader attention and driving continuous improvements in CNN architectures [8]. For instance, networks like Residual Network (ResNet), and EfficientNet introduced new architectures that further enhanced the performance of CNNs [9] [10]. In addition to CNNs, transformers using attention mechanisms have also been applied to image processing, showing excellent performance [11]. Based on these technological advancements, the MARIO project was proposed. The project contains two tasks: classifying the evolution between two pairs of 2D slices from two consecutive OCT acquisitions and predicting the evolution within 3 months of AMD on OCT 2D slices [12]. This research work aims to develop deep learning algorithms to address both tasks. For Task 1, we used a ConvNeXt-based Siamese network with multi-head attention to classify image pairs. This network structure is rarely applied in ophthalmic OCT classification tasks, highlighting the novelty of our model [13]. It is well-suited for tasks that require classification of changes between temporally related image pairs. For Task 2, a single ConvNeXt network is utilized to classify each image individually, followed by adjusting the labels of images under the same localizer through thresholding.

2 Methodology

2.1 Dataset

The dataset used in this research is sourced from the MICCAI 2024 MARIO project. The data for both tasks were acquired using the Spectralis OCT device from Heidelberg Engineering, with all data collected by ophthalmologists with over two years of work experience [12]. Table 1 shows the data distribution for the two tasks.

2.2 Data Pre-processing

To expand the dataset used for training, this study applied data augmentation strategies for both tasks. We first analyzed the data distribution in the training

Table 1. Data distribution for Task 1 and Task 2.

Category	Patients (Task 1 and 2)	Task 1 B-scan pairs	Task 2 B-scan pairs
Training	68	14,496	5,257
Validation	34	7,010	2,486
Test	64 (34 from Brest + 30 from Tlemcen)	8,341 from Brest + (unspecified from Tlemcen)	7,291 from Brest + (unspecified from Tlemcen)

set for Task 2 and found a significant imbalance. The number of images with labels 0, 1, and 2 are 1,066, 6,556, and 460, respectively. The issue of data imbalance occurs when the majority of examples are labeled as one class, known as the majority class. As a result, the classifier tends to perform well on the majority class but poorly on the minority class, where there are fewer examples [14]. To address this issue, we applied data augmentation to the training set for Task 2 before starting the training process. We used Albumentations, a flexible, fast, and easy-to-use image augmentation tool, to perform the operations listed in Table 2, generating new images and saving them for subsequent training. [15]. The number of images with categories 0, 1 and 2 are increased by 3, 1 and 5 times respectively. The image before and after data augmentation is shown in Fig. 1.

Table 2. Data augmentation operations used before training for Task 2.

Transformation	Parameters	Probability
ColorJitter	brightness = 0.3, contrast = 0.3	0.4
GaussianBlur	blur_limit = (5, 9), sigma_limit = (0.1, 5)	0.2
Affine	scale = (0.9, 1.1), translate_percent = (0.1, 0.1), rotate = 20	1.0
HorizontalFlip	–	0.4

CNNs rely on large datasets to ensure model accuracy and prevent overfitting [16]. Therefore, we also applied dynamic data augmentation during the training process for both tasks, ensuring that the model encounters different data each time. The same data augmentation operations provided by "torchvision" were applied during training for both tasks. The detailed operations are shown in Table 3. Since the data in Task 1 consists of image pairs, we applied the same transformation operations to both images in each pair. The image pair before

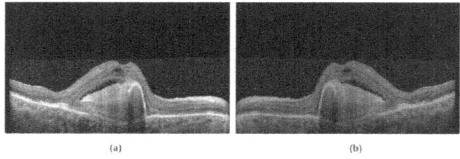

Fig. 1. The image before and after data augmentation used prior to training: (a) original image, (b) augmented image.

and after transformation are shown in Fig. 2. After the image transformations were completed, all images were resized to 224 × 224 pixels and normalized.

Table 3. Data augmentation operations used during training for both tasks.

Operation	Parameters	Probability
ColorJitter	brightness = 0.2, contrast = 0.2	0.2
GaussianBlur	kernel_size = (5, 9), sigma = (0.1, 5)	0.2
RandomAffine	degrees = 20, scale = (0.9, 1.1), translate = (0.05, 0.05)	1
RandomHorizontalFlip	–	0.3

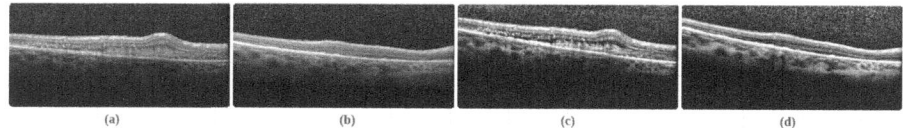

Fig. 2. The image pair before and after dynamic data augmentation used during training: (a) the first image of the pair before augmentation, (b) the second image of the pair before augmentation, (c) the first image of the pair after augmentation, (d) the second image of the pair after augmentation.

2.3 Design of Algorithms

In this research, neural networks are employed to automatically extract image features, with both tasks utilizing the ConvNeXt_large network, pre-trained on ImageNet, as the backbone. ConvNeXt is a CNN architecture proposed in 2022, which incorporates design principles from models such as the Swin Transformer, adapting and improving upon previous CNN designs. ConvNeXt adjusts the computational ratio at each stage, replaces the traditional ResNet stem with a "patchify stem", increases the network's width by using depthwise convolution, and incorporates inverted bottleneck structures along with larger convolutional kernels [17]. Since the ConvNeXt model we used is designed to accept three-channel images by default, we replicated the single-channel image into three channels to make it compatible with the ConvNeXt model. Figure 3 shows the structure of the ConvNeXt_large feature extractor used in this research.

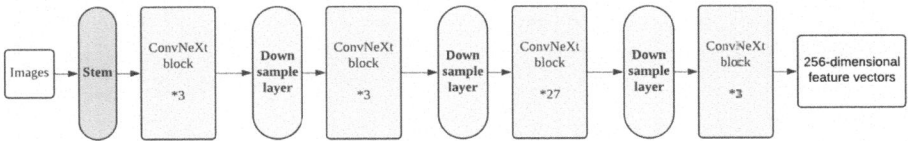

Fig. 3. The structure of the ConvNeXt_large feature extractor, adapted from [17].

Algorithm Design for Task 1. The objective of Task 1 is to observe the OCT images of a patient collected during two consecutive sessions and classify the progression in AMD. To enable the deep learning model to learn the differences between two images, we employed a Siamese network that processes both images of the pair simultaneously. The branches of a Siamese network share weights, making them particularly effective in comparative learning. These networks are able to capture similarities between instances easily, which makes them well-suited for tasks such as change detection and image matching [18]. We employed two ConvNeXt_large feature extractors as branches of the Siamese network, with each branch receiving an image generated after data augmentation, extracting its features, and producing an output of 256 dimensions. After the CNN, a multi-head attention layer is added. The multi-head attention mechanism computes a weighted average across different heads, generating a new feature vector from the input. This layer is designed to capture the important information in the input features through attention, enabling the network to focus more effectively on key features of the images [11]. Finally, after the feature vectors from both branches are combined, a linear classification layer is added. The combined vector is passed through a Rectified Linear Unit (ReLU) activation function and a linear layer to produce the final classification result. Figure 4 illustrates the general steps of the algorithm for Task 1.

Algorithm Design for Task 2. The objective of Task 2 is to predict disease progression three months later based on OCT images, only a single feature extraction network is needed. We retrained the ConvNeXt_large model, which was pre-trained on ImageNet, for feature extraction. Similarly, after feature extraction, a multi-head attention mechanism is added to capture global contextual information. Finally, the feature vectors are passed through a classification layer to obtain the predicted outcomes. Figure 5 illustrates the general procedure of the algorithm for Task 2.

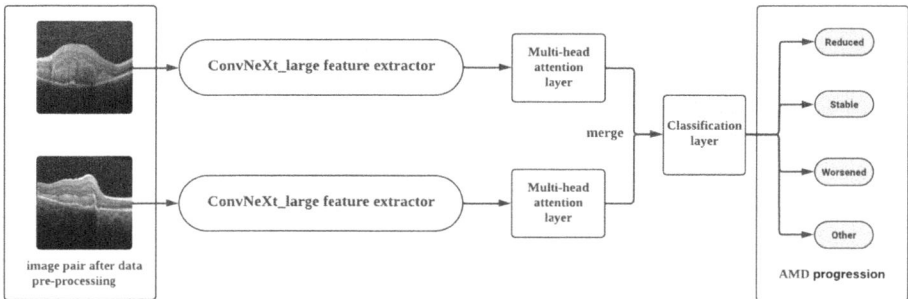

Fig. 4. The general process of the algorithm for analyzing image pair to determine AMD progression.

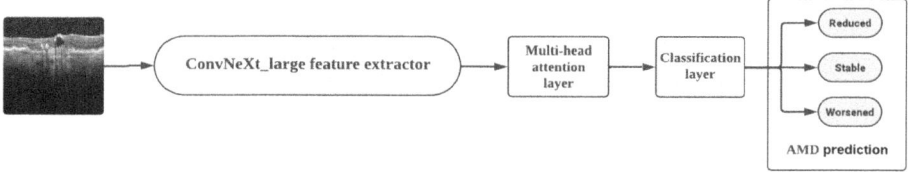

Fig. 5. The general process of algorithms for predicting AMD progression.

2.4 Post-processing and Optimization Strategy

Post-processing. In the second task, the data categories are associated with the localizer, a single localizer is applied to the entire OCT, encompassing the C-scan and multiple B-scans. Since the proposed model is trained on individual sample, we established a proportion threshold for each label by grouping the data according to the localizer. If the proportion of instances corresponding to a specific label exceeds the set threshold within a localizer, all data in that group are assigned that label. Otherwise, the original labels are maintained. We set the thresholds for labels 0, 1, and 2 at 0.4, 0.7, and 0.3, respectively.

Optimization Strategy. In both tasks, AdamW was selected as the optimizer. AdamW is an improvement over the traditional Adam optimization algorithm. While the conventional Adam algorithm directly incorporates L2 regularization into the gradient update, AdamW decouples weight decay from the optimization step relative to the loss function, preventing the excessive influence of regularization [19]. Additionally, we applied learning rate decay in both tasks, where the learning rate is reduced to 80% of its original value every two epochs. This decay mitigates oscillations of the model parameters near the optimal point and facilitates the fine-tuning process [20].

Since this is a multi-class classification task, we used cross-entropy as the loss function. In the training set for Task 1, label 1 accounts for the highest number of 9414 images, while label 3 has the lowest with 1295. Similarly, for Task 2, label 1 is the most prevalent with 6,556 images, whereas label 3 has only 460. To further address the class imbalance issue and increase the focus on minority classes, we assigned different loss weights to each label in Task 2. Specifically, we set the loss weights for classes 0, 1, and 2 to 1.5, 1, and 2, respectively.

3 Results

3.1 Results for the First Task

We evaluated the model using the validation set, with the evaluation metrics such as F1 score, RK-correlation, and specificity. The F1 score, ranging from 0 to 1, balances precision and recall, with higher scores indicating a better balance between the two metrics [21]. The RK-correlation is a generalization of the Pearson correlation coefficient, similar to the least-squares properties observed in Pearson's case [22]. Specificity measures the model ability to correctly identify negative classes. We also calculated the average of these metrics and the confusion matrix. Table 4 presents the results of the algorithm on the validation set, while Table 5 shows the confusion matrix for Task 1. The mean score suggests that data augmentation and the attention mechanism had a positive impact on the model.

Table 4. The results of the algorithm for the first task.

Multi-Head Attention	Data Augmentation	F1 Score	RK-Correlation	Specificity	Mean
No	No	0.80399	0.57348	0.87786	0.75178
Yes	No	0.80128	0.59898	0.90154	0.76727
Yes	**Yes**	**0.84508**	**0.66851**	**0.89922**	**0.80427**

3.2 Results for the Second Task

The evaluation metrics for Task 2 include F1 score, RK-correlation, specificity, and Quadratic Weighted Kappa (QWK). QWK is a statistical measure used to assess the agreement between two classification systems, with values ranging from -1 to 1, where a QWK of 1 indicates perfect agreement between the systems [23]. Table 6 and Table 7 respectively list the confusion matrix and the results after post-processing for the second task. The average results also indicate the effectiveness of data augmentation and the attention mechanism. However, the confusion matrix reflects that the impact of data imbalance remains.

Table 5. Confusion matrix of the validation set for the first task.

	Reduced	Stable	Worsened	Other
Reduced	693	294	19	27
Stable	60	4617	45	42
Worsened	35	243	408	12
Other	30	266	13	206

Table 6. Confusion matrix of the validation set for the second task.

	Reduced	Stable	Worsened
Reduced	61	290	0
Stable	101	2595	125
Worsened	57	495	98

Table 7. The results of the algorithm for the second task.

Multi-Head Attention	Data Augmentation	F1 Score	RK-Correlation	Specificity	QWK	Mean
No	No	0.67949	0.12468	0.00078	0.70812	0.37827
Yes	No	0.69571	0.11100	0.03043	0.69777	0.38373
Yes	**Yes**	**0.72057**	**0.17719**	**0.71029**	**0.14100**	**0.43726**

4 Discussions

In this MARIO project, we explored various solutions for both sub-tasks, focusing on the latest advancements in related field and actively attempting to apply new research findings to our algorithms. For both tasks, we used the advanced ConvNeXt_large as the backbone of the deep CNN and incorporated a multi-head attention mechanism. For the first task, we initially employed a single-network backbone architecture, where we calculated the difference between the features extracted by a single ConvNeXt_large feature extractor. However, as this approach did not effectively capture the distinctions, then we switched to using a Siamese network. The results show that our model could effectively compare two images in a pair and classify the progression in AMD. However, the model showed lower accuracy for minority classes, reflecting the impact of data imbalance. For Task 2, our model demonstrated the capability to predict AMD progression, but there is still a significant gap compared to the ideal algorithm. Similar to Task 1, the model demonstrated limited effectiveness in predicting minority classes. We were cautious in addressing data imbalance, aiming to avoid causing the data to deviate significantly from its real-world distribution.

This research work also has some limitations, which was conducted using an RTX 4070 GPU with 12GB of Video Random Access Memory (VRAM), which is insufficient for training large models such as EfficientNet_l2 and ConvNeXt V2. Due to time constraints, this study did not implement Test Time Augmentation (TTA), which is worth trying in future work [24]. For Task 2, we chose to classify each instance individually and then consolidate the categories within the localizer based on a threshold. This decision was made after testing some basic Multiple Instance Learning (MIL) models, which proved less effective for this task. However, this does not imply the unsuitability of all MIL models, more advanced algorithms like TransMIL remain promising and worth exploring in future studies [25].

5 Conclusions

This study accomplished both tasks proposed by the MARIO project. We actively follow advancements in this field and explore new approaches. In the first task, we implemented a Siamese network to extract features from each image in a image pair, then combined the features and get the classification results. For the second task, in addition to employing dynamic data augmentation, we also applied data augmentation before training to expand the dataset and mitigate class imbalance, followed by predictions using a ConvNeXt_large-based network. Both tasks incorporated a multi-head attention mechanism, enhancing the focus on important features. The algorithms demonstrated strong performance on the validation set, achieving an average rank of third place for both tasks during the development phase. Given the rapid advancements in this field, there remains substantial potential for further enhancement. Future research is expected to actively explore emerging technologies, continually enhancing the power of deep learning in medical image processing.

References

1. Mitchell, P., Liew, G., Gopinath, B., Wong, T.Y.: Age-related macular degeneration. Lancet **392**(10153), 1147–1159 (2018)
2. Jonas, J.B., Cheung, C.M.G., Panda-Jonas, S.: Updates on the epidemiology of age-related macular degeneration. Asia-Pac. J. Ophthalmol. **6**(6), 493–497 (2017)
3. Fleckenstein, M., et al.: Age-related macular degeneration. Nat. Rev. Dis. Primers. **7**, 31 (2021)
4. Keane, P.A., Patel, P.J., Liakopoulos, S., Heussen, F.M., Sadda, S.R., Tufail, A.: Evaluation of age-related macular degeneration with optical coherence tomography. Surv. Ophthalmol. **57**(5), 389–414 (2012)
5. Elsharkawy, M., et al.: Role of optical coherence tomography imaging in predicting progression of age-related macular disease: a survey. Diagnostics **11**(12), 2313 (2021)
6. Karampelas, M., Malamos, P., Petrou, P., Georgalas, I., Papaconstantinou, D., Brouzas, D.: Retinal pigment epithelial detachment in age-related macular degeneration. Ophthalmol. Ther. **9**, 739–756 (2020)
7. Srivastava, O., Tennant, M., Grewal, P., Rubin, U., Seamone, M.: Artificial intelligence and machine learning in ophthalmology: a review. Indian J. Ophthalmol. **71**(1), 11–17 (2023)
8. Krizhevsky, A., Sutskever, I., Hinton, G.E.: ImageNet classification with deep convolutional neural networks. Commun. ACM **60**(6), 84–90 (2017)
9. He, K., Zhang, X., Ren, S., Sun, J.: Deep residual learning for image recognition. In: 2016 IEEE Conference on Computer Vision and Pattern Recognition (CVPR), pp. 770–778. IEEE, Las Vegas (2016)
10. Tan, M., Le, Q.V.: EfficientNet: rethinking model scaling for convolutional neural networks. ArXiv abs/1905.11946 (2019)
11. Dosovitskiy, A., et al.: An image is worth 16×16 words: transformers for image recognition at scale. ArXiv abs/2010.11929 (2020)
12. Zeghlache, R., et al.: Monitoring age-related macular degeneration progression in optical coherence tomography. In: 27th International Conference on Medical Image Computing and Computer Assisted Intervention (MICCAI 2024). Zenodo (2024)

13. Abd El-Khalek, A.A., et al.: A comprehensive review of AI diagnosis strategies for age-related macular degeneration (AMD). Bioengineering **11**(7), 711 (2024)
14. Singh, A., Purohit, A.: A survey on methods for solving data imbalance problem for classification. Int. J. Comput. Appl. **127**, 37–41 (2015)
15. Buslaev, A., Iglovikov, V.I., Khvedchenya, E., Parinov, A., Druzhinin, M., Kalinin, A.A.: Albumentations: fast and flexible image augmentations. Information **11**(2), 125 (2020)
16. Shorten, C., Khoshgoftaar, T.M.: A survey on image data augmentation for deep learning. J. Big Data **6**, 1–48 (2019)
17. Liu, Z., Mao, H., Wu, C.-Y., Feichtenhofer, C., Darrell, T., Xie, S.: A ConvNet for the 2020s. In: 2022 IEEE/CVF Conference on Computer Vision and Pattern Recognition (CVPR), pp. 11966–11976. IEEE, New Orleans (2022)
18. Li, Y., Chen, C.L.P, Zhang, T.: A survey on siamese network: methodologies, applications, and opportunities. IEEE Trans. Artif. Intell. **3**, 994–1014 (2022)
19. Loshchilov, I., Hutter, F.: Decoupled weight decay regularization. In: International Conference on Learning Representations (2017)
20. You, K., Long, M., Wang, J., Jordan, M.I.: How does learning rate decay help modern neural networks. ArXiv abs/1908.01878 (2019)
21. Opitz, J., Burst, S.: Macro F1 and macro F1. ArXiv abs/1911.03347 (2019)
22. Gorodkin, J.: Comparing two K-category assignments by a K-category correlation coefficient. Comput. Biol. Chem. **28**, 367–374 (2004)
23. Cohen, J.: Weighted kappa: nominal scale agreement with provision for scaled disagreement or partial credit. Psychol. Bull. **70**(4), 213–220 (1968)
24. Shanmugam, D., Blalock, D., Balakrishnan, G., Guttag, J.: Better aggregation in test-time augmentation. In: Proceedings of the IEEE/CVF International Conference on Computer Vision (ICCV), pp. 1194–1203. IEEE, Montreal (2021)
25. Shao, Z., et al.: TransMIL: transformer based correlated multiple instance learning for whole slide image classification. ArXiv abs/2106.00908 (2021)

Leveraging MaxVit on Fused OCT Scan Pairs for Age-Related Macular Degeneration Evolution Assessment

Yosuke Yamagishi[✉][iD]

The University of Tokyo, Tokyo 113-8655, Japan
`yamagishi-yosuke0115@g.ecc.u-tokyo.ac.jp`

Abstract. This paper presents our approach to the MARIO (Monitoring Age-related Macular Degeneration Progression In Optical Coherence Tomography) challenge at MICCAI 2024, focusing on automated analysis of Age-related Macular Degeneration (AMD) evolution using Optical Coherence Tomography (OCT) images. For Task 1, we propose a novel method that fuses consecutive OCT scan pairs, utilizing a MaxVit Tiny architecture to classify AMD progression. This approach mimics clinicians' side-by-side comparison technique. In the preliminary phase, our model ranked first among 20 teams across all evaluation metrics for Task 1, achieving an F1 score of 0.863, Matthews Rk-correlation coefficient (RkC) of 0.709, and Specificity of 0.913. For Task 2, predicting AMD progression over three months, we combined EfficientNet V2 S for OCT image processing with a Multi-Layer Perceptron for handling patient metadata. While our Task 2 solution ranked competitively (3rd in F1, 1st in QWK among 16 teams), overall scores were low (F1: 0.685, RkC: 0.111, QWK: 0.231, Specificity: 0.688), highlighting challenges in long-term progression prediction. Our Task 1 solution demonstrates significant potential for improving AMD monitoring efficiency in clinical settings, while Task 2 results underscore areas needing further research. This work contributes to enhancing AMD management, with implications for improved patient care in ophthalmology.

Keywords: Age-related Macular Degeneration (AMD) · Optical Coherence Tomography (OCT) · Deep Learning · MaxVit · Disease Progression · MICCAI Challenge

1 Introduction

Age-related Macular Degeneration (AMD) is a progressive eye condition that affects millions of people worldwide, particularly in aging populations [4]. It is a leading cause of vision loss and significantly impacts the quality of life of affected individuals. Monitoring the progression of AMD is crucial for timely intervention and effective treatment planning [1]. Optical Coherence Tomography (OCT) has emerged as a key imaging modality in ophthalmology [5], providing high-resolution, cross-sectional images of the retina that are invaluable for diagnosing and monitoring AMD.

The MARIO (Monitoring Age-related Macular Degeneration Progression In Optical Coherence Tomography) challenge, part of MICCAI Challenges 2024, aims to advance the automated analysis of AMD progression using OCT images [13]. This challenge addresses two critical tasks in AMD management: the classification of disease evolution between consecutive OCT scans (Task 1) and the prediction of disease progression over a three-month period (Task 2).

Our research develops innovative deep learning solutions for both tasks in the MARIO challenge, focusing primarily on Task 1. The method of comparing medical images side by side for visual assessment is commonly employed to detect changes in lesions. This process, while effective, is time-consuming and subject to inter-observer variability. We propose a novel approach that mirrors clinicians' side-by-side comparison method by fusing pairs of consecutive OCT scans. Using the MaxVit Tiny architecture [11], we classify AMD progression based solely on image data. This method captures subtle retinal changes indicative of disease evolution by effectively processing both local and global image features.

For Task 2, which involves predicting AMD progression over a longer time frame, we developed a hybrid approach. This solution combines an EfficientNet V2 S [10] for processing OCT images with a Multi-Layer Perceptron (MLP) for handling patient metadata such as age and gender. By integrating image features with relevant clinical data, we aim to provide a more comprehensive prediction of disease progression over a three-month period.

2 Task and Dataset

2.1 Task 1: Evolution Classification

Task 1 aims to classify the evolution of AMD between consecutive OCT examinations, assessing whether the condition has improved, remained stable, or deteriorated.

The dataset comprises OCT scans from 136 vascular AMD patients, acquired using a Spectralis OCT device at Brest University Hospital, France. It is divided into 68 training, 34 validation, and 34 test cases. Each case includes multiple 3-D volumes (C-scans) and 2-D infrared localizer images.

For annotation, consecutive B-scans were viewed jointly on the same screen, with the older examination at the top and the newer at the bottom. The task involves a four-class evolution assessment between consecutive examinations:

- REDUCED
- STABLE
- WORSENED
- OTHER

Annotations were performed by experienced ophthalmologists, with an inter-annotator agreement of approximately 75%. The validation set was used for the competition's preliminary phase, while the undisclosed test set determined final rankings.

2.2 Task 2: Progression Prediction

Task 2 focuses on predicting AMD progression over a three-month period, aiming to forecast whether a patient's condition will improve, remain stable, or worsen based on current OCT scans and patient metadata.

The dataset structure and acquisition methods are identical to Task 1, using the same split of 68 training, 34 validation, and 34 test cases. Unlike Task 1, which assesses changes between consecutive examinations, this task predicts future disease evolution and incorporates additional patient metadata such as age, gender, and number of visits.

The task involves a three-class classification system:

- REDUCE
- STABLE
- INCREASE

Notably, all examination dates were masked to prevent predictions based on treatment timing, ensuring the focus remains on clinical features and patient characteristics. The annotation process and inter-annotator agreement remain consistent with Task 1.

3 Method

3.1 Task 1

Figure 1 illustrates the overview of our approach for Task 1. The system takes two consecutive OCT scans as input, resizes and concatenates them vertically into a single 512×512 pixel image, which is then fed into a MaxViT Tiny model for classification into four categories.

Data Preprocessing. We implemented the following preprocessing steps to effectively handle consecutive OCT images:

1. Each OCT image was resized to 512×256 pixels to ensure uniformity
2. Two consecutive resized OCT images were vertically concatenated, resulting in a single 512×512 pixel image
3. This fusing approach creates a side-by-side representation of consecutive scans

This preprocessing method emulates the side-by-side comparison typically employed by clinicians in actual diagnostic processes. By presenting the images in this manner, we aim to enhance the model's ability to learn temporal changes between consecutive OCT scans.

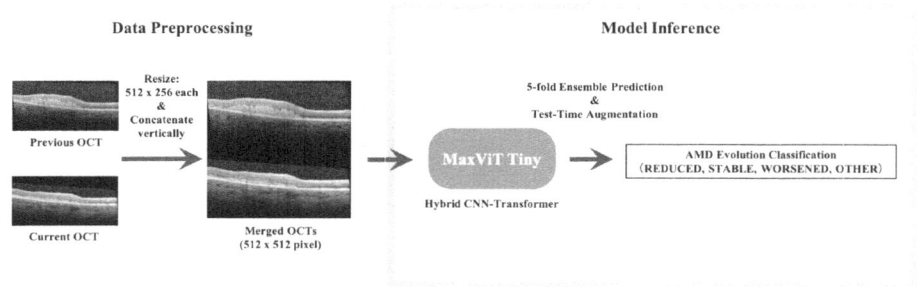

Fig. 1. Overview of our approach for Task 1. The system processes consecutive OCT scans through two main stages: (1) Data Preprocessing, where pairs of OCT images (512 × 256 pixels each) are vertically concatenated to create a single 512 × 512 pixel merged image, mimicking clinical side-by-side comparison; and (2) Model Inference, utilizing MaxViT Tiny with 5-fold ensemble prediction and test-time augmentation for final classification into four AMD evolution categories (REDUCED, STABLE, WORSENED, OTHER).

Model Architecture. For our model architecture, we adopted MaxViT Tiny, a hybrid model that combines the strengths of both Convolutional Neural Networks (CNNs) and Transformers. This choice was motivated by the specific requirements of our task and the complementary nature of these architectural paradigms.

CNNs are known for their ability to capture local textures and patterns effectively [2]. However, they may struggle with extracting global features across an entire image. On the other hand, Transformer-based models excel at capturing long-range dependencies and global features, making them particularly suited for tasks that require understanding the overall context of an image [2]. Moreover, a key feature of MaxVit is its ability to acquire global information more efficiently compared to other high-performance models like Swin Transformer [6] or ConvNeXt [7]. Therefore, we considered it optimal for this task, as it is expected to extract more context-based information from two fused images.

For Task 1, we also employed EfficientNet V2 S for performance comparison purposes.

MaxViT was trained using weights pretrained on ImageNet-1K [9]. For EfficientNetV2, weights that were initially pretrained on ImageNet-22K [8] and then fine-tuned on ImageNet-1K were used.

Training. The training process utilized a StratifiedGroupKFold cross-validation strategy with 5 folds to ensure robust model evaluation. This approach maintained patient-wise grouping to prevent data leakage between folds, while also stratifying based on target labels to preserve the class distribution across folds. For each fold, we employed the following procedure:

1. Data Augmentation: We applied various augmentation techniques using the albumentations library to enhance model generalization. These included:
   - Random resized crop (scale: 0.85 to 1.0)
   - Random horizontal flip (probability: 0.5)
   - Random rotation (up to 15°)
   - Coarse dropout (up to 5 holes, each covering up to 5% of the image)
2. Model Training: We used the Adam optimizer with a learning rate of 1e−4 and weight decay of 1e−6. The learning rate was adjusted using a cosine schedule. Training was conducted for 5 epochs with a batch size of 16.
3. Loss Function: Cross-Entropy Loss was used as the criterion for model optimization.
4. Evaluation Metrics: During training, we monitored multiple metrics, including the F1-score, Matthews Rk-correlation coefficient (RkC), and Specificity, which were adopted as evaluation metrics for the shared task. The average of these three metrics was used to select the best model checkpoint.
5. Model Selection: The model checkpoint achieving the highest average score across the evaluation metrics was saved for each fold.

Inference. For inference, we employed ensemble prediction and test-time augmentation (TTA) strategies:

1. Ensemble Prediction: Predictions were made using an ensemble of models from all 5 trained folds. The final prediction was obtained by averaging the outputs of these 5 models.
2. TTA: For each image, predictions were made on both the original and horizontally flipped versions, with the results averaged.

3.2 Task 2

Figure 2 illstrates the overview of our approach for Task 2. The key differences from Task 1 are the fusing of the Localizer images in addition to the OCT images to form a single image, and the adoption of a multimodal model that incorporates metadata such as the patient's age and gender, given the challenge of predicting prognosis using image data alone.

Data Preprocessing. Task 2 preprocessing differs from Task 1 in the following ways:

- OCT Image Preprocessing:
  - OCT image was resized to 512×256 pixels.
  - The resized OCT image was vertically concatenated with its corresponding Localizer image (also resized to 512×256 pixel), resulting in a single 512×512 pixel image.
- Localizer Image Inclusion:
  - Localizer images were included to provide additional context at the study level, as labels were assigned per study rather than per individual image.

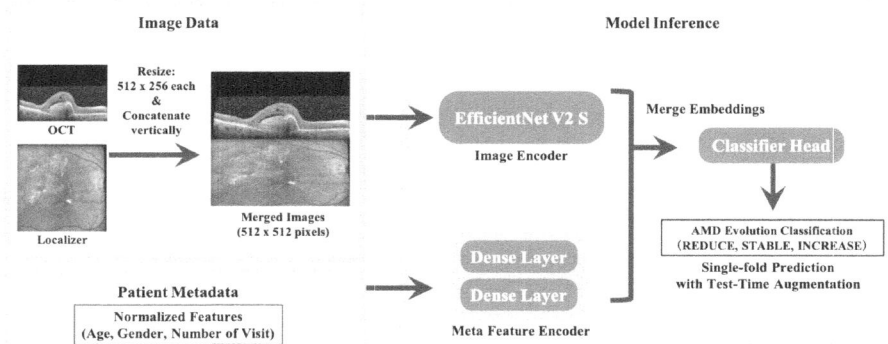

Fig. 2. Overview of our approach for Task 2. The system combines two parallel data streams: (1) Image Data processing, where OCT and Localizer images are resized and vertically concatenated into a 512 × 512 pixel merged image, processed through EfficientNet V2 S; and (2) Patient Metadata processing, where normalized demographic features are encoded through two Dense Layers. Both streams' embeddings are merged and passed through a Classifier Head for final three-class prediction (REDUCE, STABLE, INCREASE), using single-fold prediction with test-time augmentation.

– Patient Metadata Preprocessing:
  • Age: Normalized by dividing by 100.
  • Gender: Binarized (0 for female, 1 for male).
  • Number of visits: Normalized by dividing by 10.

Model Architecture. For Task 2, we employed a multimodal architecture:

1. Image Encoder: EfficientNet V2 S
2. Meta Feature Encoder: two Dense layers (first layer expands features to 128 dimensions, second layer reduces to 64 dimensions)
3. Feature Fusion and Classification: Concatenated features processed through a Dense layer

Training. The training process is similar to Task 1, with the following difference:

– Model Selection: The final model was selected based on the mean of the evaluation metrics (F1 score, RkC, QWK, and Specificity) across the 5-fold cross-validation.

Other training settings (data augmentation, optimizer, learning rate, epochs, batch size, loss function) remain the same as in Task 1.

Additionally, for comparison purposes, we also trained a model that only uses image data without incorporating metadata.

Inference. The inference pipeline for Task 2 differed from Task 1 in that it used only a single fold out of the 5-fold cross-validation for model construction and inference. However, it maintained the same TTA strategy as Task 1

3.3 Code Availability

The code used for this study is publicly available on GitHub at the following repository: https://github.com/yamagishi0824/MARIO24-MaxVit-Fused.

4 Results

The MARIO challenge was conducted on the Codabench platform [12]. The challenge consisted of two phases: a preliminary phase and a final phase. We present the results from both phases separately to provide a comprehensive evaluation of our method's performance.

4.1 Task 1

Our MaxVit Tiny-based approach demonstrated exceptional performance during the preliminary phase of the MARIO challenge, which involved 20 participating teams. Our method achieved the highest scores across all three evaluation metrics (F1 Score, RkC, and Specificity) among the competing teams.

To further validate our method, we compared the performance of our MaxVit Tiny-based approach with that of EfficientNet V2 S, another state-of-the-art architecture. Table 1 presents a side-by-side comparison of the two models across all evaluation metrics for the preliminary phase.

Table 1. Performance Comparison of MaxVit and EfficientNet V2 S for Task 1 (Preliminary Phase)

Model	F1	RkC	Specificity
MaxVit Tiny	**0.863**	**0.709**	**0.913**
EfficientNet V2 S	0.837	0.657	0.906

As evident from Table 1, our MaxVit Tiny-based approach consistently outperformed the EfficientNet V2 S model across all evaluation metrics during the preliminary phase. This consistent superiority can be attributed to MaxVit's ability to effectively capture both local and global features in the fused OCT scan pairs, which is crucial for accurate progression assessment.

4.2 Task 2

For Task 2 of the MARIO challenge, our approach also demonstrated strong performance among the 16 participating teams. Our method achieved competitive scores across all four evaluation metrics.

Table 2. Comparison of Performance with and without Metadata (Scores for the model with metadata include leaderboard rankings in parentheses)

Metadata	F1	RkC	Specificity	QWK
Yes	0.685 (3)	**0.111** (5)	**0.688** (5)	**0.231** (1)
No	**0.727**	0.081	0.687	−0.126

As shown in Table 2, our approach achieved the top rank in the QWK and third place in the F1 Score. However, it is crucial to note that the overall scores across all metrics are notably low, raising important considerations about the effectiveness of current approaches in this challenging task. Our highest score was in Specificity (0.688), while our top-ranked QWK score was only 0.231. The RkC score was particularly low at 0.111.

By incorporating metadata, improvements were observed in all evaluation metrics except for the F1 score. Without metadata, the QWK value is negative, indicating that the model likely fails to make any meaningful predictions.

4.3 Final Results

In the final evaluation phase of the MARIO Challenge, our method demonstrated strong performance in Task 1, achieving second place among 12 participating teams. The model achieved consistently high scores across all evaluation metrics (F1 Score: 0.855, RkC: 0.676, Specificity: 0.914).

For Task 2, our approach faced significant challenges, particularly in capturing the progression patterns. While the F1 Score (0.810) and Specificity (0.667) were moderate, the RkC (0.018) and QWK (0.073) scores showed substantial degradation from the preliminary phase. Notably, the RkC ranked lowest among the 12 participating teams, while the QWK dropped to tenth place, indicating significant difficulties in accurately ordering and predicting the progression of the disease.

5 Discussion

Our study presents a novel approach to automated AMD progression assessment, addressing the challenges posed in the MARIO competition at MICCAI 2024. The results demonstrate the potential of our methods, particularly for Task 1.

For Task 1, our approach of using MaxVit architecture on fused, paired consecutive OCT images proved highly effective. This method successfully mimicked the side-by-side comparison performed by clinicians, leveraging both local and global features of the images. While our preliminary experiments included attempts to incorporate localizer images through fusion, this did not yield improvements in validation scores. However, more sophisticated approaches that consider the spatial context and precise positioning information from localizer images might prove beneficial, more closely mimicking the way clinicians integrate this information in their diagnostic process. Achieving the top rank among 20 competing teams in all evaluation categories further validates the promise of this methodology for clinical application in AMD monitoring.

Task 2 results reveal significant challenges in long-term progression prediction. While our preliminary results were promising, the final evaluation phase showed substantial performance degradation, particularly in progression ordering metrics. The RkC scored lowest among 12 teams, and the QWK dropped to tenth place, suggesting fundamental difficulties in capturing disease progression patterns. While our results demonstrate the benefit of including metadata, we have not yet conducted a detailed analysis of the relative importance of each contextual feature (age, gender, and number of visits). A key reason for this poor performance is that identical labels are assigned to entire studies rather than individual images, introducing noise that hampers the model's learning. We hypothesize that multiple instance learning could address this issue [3].

This study has several limitations. One is the limited number of model architectures that were tested. Exploring various backbone architectures could potentially lead to further performance improvements. Additionally, in Task 1, fusing was performed by mimicking a manual approach, but various other methods could be explored. For example, stacking in the channel direction or inputting the two images separately into the encoder for prediction are some of the variations that could be considered.

In conclusion, our research demonstrates significant progress in automating AMD progression assessment, particularly for short-term changes between consecutive OCT scans. The success of our MaxVit-based approach in Task 1 offers a promising direction for clinical application. However, the challenges encountered in Task 2 underscore the complexity of long-term AMD progression prediction and highlight the need for further research in this area. Future work should focus on addressing these limitations, possibly through more advanced machine learning techniques, detailed analysis of metadata importance, or by incorporating additional clinical data beyond OCT imaging.

Acknowledgments. This research received no specific grant from any funding agency in the public, commercial, or not-for-profit sectors.

Disclosure of Interests. There is no conflict of interest to disclose.

References

1. Freund, K.B., et al.: Treat-and-extend regimens with anti-VEGF agents in retinal diseases: a literature review and consensus recommendations. Retina **35**(8), 1489–1506 (2015)
2. Han, K., et al.: A survey on vision transformer. IEEE Trans. Pattern Anal. Mach. Intell. **45**(1), 87–110 (2022)
3. Ilse, M., Tomczak, J., Welling, M.: Attention-based deep multiple instance learning. In: International Conference on Machine Learning, pp. 2127–2136. PMLR (2018)
4. Jonas, J.B., Cheung, C., Panda-Jonas, S.: Updates on the epidemiology of age-related macular degeneration. Asia-Pac. J. Ophthalmol. **6**(6), 493–497 (2017)
5. Keane, P.A., Patel, P.J., Liakopoulos, S., Heussen, F.M., Sadda, S.R., Tufail, A.: Evaluation of age-related macular degeneration with optical coherence tomography. Surv. Ophthalmol. **57**(5), 389–414 (2012)
6. Liu, Z., et al.: Swin transformer: hierarchical vision transformer using shifted windows. In: Proceedings of the IEEE/CVF International Conference on Computer Vision, pp. 10012–10022 (2021)
7. Liu, Z., Mao, H., Wu, C.Y., Feichtenhofer, C., Darrell, T., Xie, S.: A convnet for the 2020s. In: Proceedings of the IEEE/CVF Conference on Computer Vision and Pattern Recognition, pp. 11976–11986 (2022)
8. Ridnik, T., Ben-Baruch, E., Noy, A., Zelnik-Manor, L.: ImageNet-21k pretraining for the masses. arXiv preprint arXiv:2104.10972 (2021)
9. Russakovsky, O., et al.: ImageNet large scale visual recognition challenge. Int. J. Comput. Vision **115**, 211–252 (2015)
10. Tan, M., Le, Q.: EfficientNetv2: smaller models and faster training. In: International Conference on Machine Learning, pp. 10096–10106. PMLR (2021)
11. Tu, Z., et al.: MaxViT: multi-axis vision transformer. In: European Conference on Computer Vision, pp. 459–479. Springer (2022)
12. Xu, Z., et al.: Codabench: flexible, easy-to-use, and reproducible meta-benchmark platform. Patterns **3**(7) (2022)
13. Youven, Z.: Mario-challenge-MICCAI-2024 (2024). https://github.com/YouvenZ/MARIO-Challenge-MICCAI-2024. Accessed 28 Aug 2024

Patch Progression Masked Autoencoder with Fusion CNN Network for Classifying Evolution Between Two Pairs of 2D OCT Slices

Philippe Zhang[1,2,3(✉)], Weili Jiang[4], Yihao Li[5], Jing Zhang[1,2], Sarah Matta[1,2], Yubo Tan[6], Hui Lin[7], Haoshen Wang[8], Jiangtian Pan[9], Hui Xu[10], Laurent Borderie[3], Alexandre Le Guilcher[3], Béatrice Cochener[1], Chubin Ou[11], Gwenolé Quellec[1], and Mathieu Lamard[1,2]

[1] LaTIM UMR 1101, Inserm, Brest, France
pzhang.wj88@gmail.com
[2] University of Western Brittany, Brest, France
[3] Evolucare Technologies, Villers-Bretonneux, France
[4] College of Computer Science, Sichuan University, Chengdu, China
[5] United Imaging Healthcare, Shanghai, China
[6] University of Electronic Science and Technology of China, Chengdu, China
[7] Northwestern University, Evanston, IL, USA
[8] ShanghaiTech University, Shanghai, China
[9] Wuhan University, Wuhan, China
[10] MD Anderson Cancer Center, Houston, USA
[11] Guangdong General Hospital, Guangzhou, Guangdong, China

Abstract. Age-related Macular Degeneration (AMD) is a prevalent eye condition affecting visual acuity. Anti-vascular endothelial growth factor (anti-VEGF) treatments have been effective in slowing the progression of neovascular AMD, with better outcomes achieved through timely diagnosis and consistent monitoring. Tracking the progression of neovascular activity in OCT scans of patients with exudative AMD allows for the development of more personalized and effective treatment plans. This was the focus of the Monitoring Age-related Macular Degeneration Progression In Optical Coherence Tomography (MARIO), in which we participated. In Task 1, which involved classifying the evolution between two pairs of 2D slices from consecutive OCT acquisitions, we employed a fusion CNN network with model ensembling to further enhance the model's performance. For Task 2, which focused on predicting progression over the next three months based on current exam data, we proposed the Patch Progression Masked Autoencoder that generates an OCT for the next exam and then classifies the evolution between the current OCT and the one generated using our solution from Task 1. The results we achieved allowed us to place in the **Top 10** for both tasks. Some team members are part of the same organization as the challenge organizers; therefore, we are not eligible to compete for the prize. The code is available at https://github.com/pzhangwj/mario_challenge_code.

Keywords: Age-related Macular Degeneration · Optical coherence tomography · Deep learning · Model ensemble

1 Introduction

Age-related Macular Degeneration (AMD) is a common condition affecting vision in individuals over 65 [9]. Since 2007, anti-VEGF treatments [14] have been vital for managing neovascular AMD, relying on early diagnosis and consistent monitoring [5,13]. Developing effective computer-assisted diagnostic tools for AMD remains a key research focus [11]. Optical coherence tomography (OCT) [10] is crucial for diagnosing and monitoring AMD, guiding treatment decisions.

Deep learning (DL) has transformed medical imaging, significantly advancing the detection of eye conditions like AMD [2] and Diabetic Retinopathy (DR) [3]. Notable DL models for time-series prediction include Long Short-Term Memory (LSTM) [7] and Vision Transformer (ViT) [4], which effectively capture temporal and spatial patterns, respectively, enhancing disease progression prediction [1,8].

The MARIO Challenge[1] aims to evaluate algorithms for detecting changes in neovascular activity from OCT scans in AMD patients. The challenge includes two tasks: analyzing consecutive OCT B-scan images to classify changes and predicting disease progression from a single scan. Our team achieved top-10 results by employing the Fusion CNN Network and a Vision Transformer-based method for these tasks.

2 Materials and Methods

2.1 Dataset Description

The MARIO dataset was collected to monitor the progression of AMD in 136 patients from Europe. For disease monitoring, each patient undergoes an examination using the Spectralis OCT device (Heidelberg Engineering) with follow-up option on. Each examination includes a set of successive 2-D OCT slices (B-scans) that form a 3-D OCT volume, a 2-D infrared image (localizer) corresponding to each 3-D volume, and associated data such as age, sex, visit number, and eye laterality. The dataset is divided as follows for both tasks: 68 patients for training, 34 for validation, and 34 for testing. To maintain the challenge's integrity and fairness, the validation and test datasets are processed by the organizers to ensure an unbiased evaluation. The test dataset used for final ranking is securely held and not released to participants.

The first Task (Classify evolution between two pairs of 2-D slices from two consecutive 2-D OCT acquisitions) is focused on a four-category image classification. The categories are as follows:

- Reduced (0): The condition has been classified as reduced, indicating that it has either been eliminated or is persistently reduced.

[1] https://youvenz.github.io/MARIO_challenge.github.io/.

- Stable (1): The condition is considered stable, meaning it is either inactive or persistently stable.
- Worsened (2): The condition has worsened, signifying that it is either persistently worsened or has newly appeared.
- Other (3): The condition falls into the "other" category, which includes cases that are uninterpretable or have both appeared and then been eliminated.

The second task (Prediction of AMD progression within 3 months using OCT 2D slices) focuses on classifying the progression into the same categories as in Task 1, excluding the "Other" category (3).

The data annotation was meticulously performed by ophthalmologists specializing in retina care, each with at least two years of experience in monitoring vascular AMD patients. Initially, seven labels were used, but this was simplified to four categories for Task 1 and three for Task 2, as described above. For the training and validation cases, a single ophthalmologist handled the annotation. In contrast, two ophthalmologists independently annotated the test cases, with the agreement between their annotations serving as a baseline for evaluating algorithm performance.

2.2 Data Processing

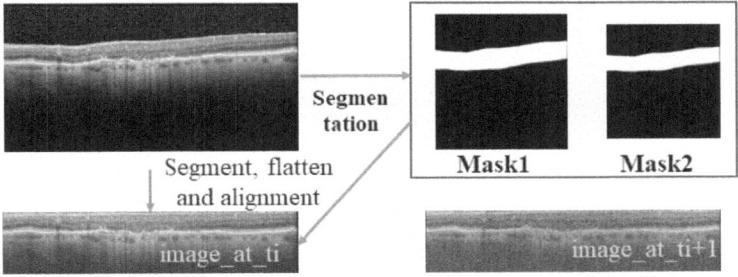

Fig. 1. Optical Coherence Tomography Image Preprocessing

OCT slices often contain significant noise and irrelevant information. While the Follow-Up option allows for precise horizontal realignment of image pairs, vertical (depth) alignment remains inconsistent. To address this and improve comparability between image pairs, we applied Optical Coherence Tomography Image Preprocessing(OCTIP[2]) [15], developed by LaTIM Labs (https://latim.univ-brest.fr/). Several Feature Pyramid Network (FPN) architectures from the EfficientNet family were trained for the segmentation task, and among them, FPN-EfficientNet-B6 and FPN-EfficientNet-B7 achieved the best performance.

---
[2] https://github.com/leto-atreides-2/octip.

We used these two top-performing models to generate segmentation masks and applied the median of their outputs to ensure a robust prediction, minimizing the impact of potential classification errors or over-segmentation from a single model. Once the segmented regions are extracted, we align the upper boundary of the retina (inner limiting membrane) with the top of the image, effectively "flattening" the retina and realigning the images along the depth axis. Additionally, by eliminating the vitreous, it enables us to zoom in on the retina, providing more detailed information to the neural network, ensuring more accurate and consistent analysis. The processing flow is illustrated in Fig. 1.

2.3 Task 1

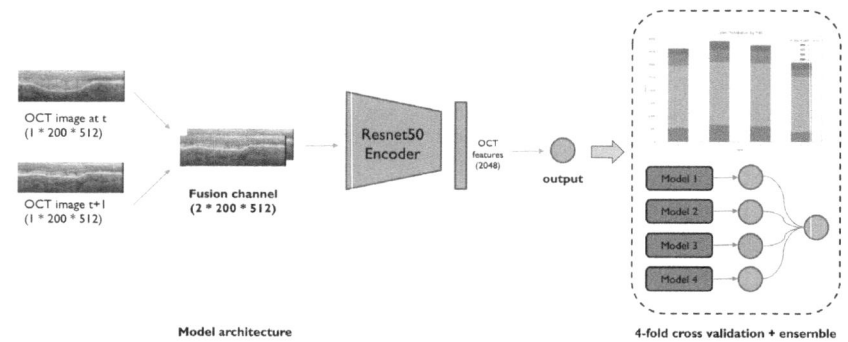

Fig. 2. Early Fusion CNN Network

In Task 1, we explored two fusion methods [12]: Early Fusion and Late Fusion networks. During the experimentation phase, we tested various backbones, and ResNet50 showed the best performance.

Early Fusion Network. Illustrated in Fig. 2, this architecture concatenates OCT images along the channel dimension. Each OCT image, with a size of $1 \times 200 \times 512$, is combined to form a fused input of size $2 \times 200 \times 512$. This fused input is passed through a ResNet50 encoder pre-trained on ImageNet, which extracts feature maps of size 2048. The extracted features are then processed through fully connected layers to perform the final classification task.

Late Fusion Network. As shown in Fig. 3, this architecture fuses image features from two different time points. The input consists of OCT images at time t and $t+1$ (with dimensions of $3 \times 200 \times 512$). Since OCT images are originally grayscale, they are repeated three times to simulate RGB format input. The OCT images from two different time points are fed into a ResNet50 encoder to extract 2×2048-dimensional features. The features extracted from the two time points are then concatenated to form a 4096-dimensional fusion feature vector (2048+2048), which is then passed through a fully connected layer to produce the classification output.

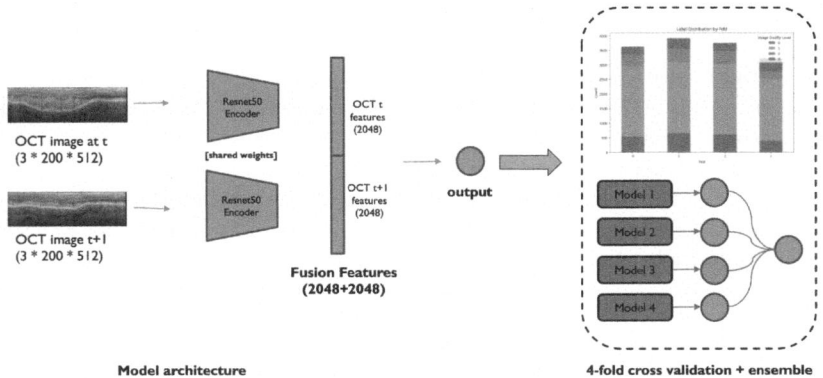

Fig. 3. Late Fusion CNN Network

Additionally, we adopted a 4-fold cross-validation strategy, training four models on different folds (Model 1 to Model 4) and combining their results through ensembling to further improve the model's robustness and generalization [16].

2.4 Task 2

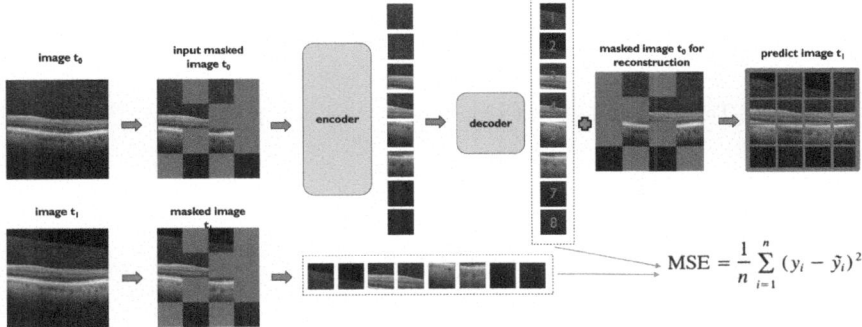

Fig. 4. Vision Transformer Patch Progression Masked AutoEncoder

Task 2 focuses on classifying disease progression based on a single examination. To address this, we propose a novel method called Patch Progression Masked Autoencoder (PPMAE), inspired by the Masked Autoencoder (MAE) [6]. While

Task 1 provides two scans to classify disease progression, Task 2 only supplies one scan. Therefore, we train PPMAE on the Task 1 dataset to predict a future OCT image based on the current scan. By generating this predicted image, we can apply the classification model from Task 1 to assess disease progression, effectively using both the current and predicted future scans to determine progression based on two scans.

Patch Progression Masked Autoencoder. Illustrated in Fig. 4, we mask 75% of the baseline OCT image at time t_0, similar to the traditional MAE method. However, instead of reconstructing the same image from the unmasked part, PPMAE predicts the patches of the follow-up image at time t_1. The model learns to use the masked patches of the t_0 image to predict the corresponding progression patches from the t_1 image. This strategy allows the model to capture temporal changes between the patches, learning the disease progression over time.

The predicted patches are then aligned with the unmasked regions of the image t_0, resulting in a reconstruction of the complete future OCT image that corresponds to the image several months later. The reconstruction error is calculated using the Mean Squared Error (MSE) between the predicted patches and the true patches from the follow-up image. This approach not only enables the model to predict the future state of the disease but also helps in improving classification performance for progressive cases by leveraging temporal information.

Finally, we leverage our classification model from Task 1 to perform the disease progression classification, using the reconstructed t_1 image along with t_0 to complete the task.

2.5 Implementation Details

The implementation details for our experiments are summarized in Table 1. All our models were trained on a NVIDIA A6000 with 48 GB of memory.

Table 1. Implementation details used in experiments.

Implementation	Task1	Task2 Finetuning	PPMAE
Input size	512 × 200 pixels		224 × 224 pixels
Backbone	ResNet50		MAE ViT Large
Library	timm		
Pretrained weights	Imagenet		Imagenet MAE
Loss	CrossEntropyLoss		MSELoss
Optimizer	AdamW		
Learning rate	1e−4 (w/o scheduler)	1e−5 (w/o scheduler)	1e−5 (w/o scheduler)
Augmentation	RandomHorizontalFlip, RandomVerticalFlip, RandomRotation, ColorJitter, RandomPerspective, GaussianBlur		RandomResizedCrop, RandomHorizontalFlip
Batch size	128		
Epochs	150	1	100
Train/Val split	0.75:0.25	1:0	0.75:0.25
Metric	Accuracy, F1, Specificity, RkC	Accuracy, F1, RkC, Specificity, QWK	MSE

3 Results

Table 2 shows OCTIP improves performance for both Early Fusion and Late Fusion methods in Task 1 of the MARIO Challenge. OCTIP enhanced F1 scores and rank correlation, particularly benefiting both Late Fusion method.

Table 2. Preliminary results on MARIO Challenge - Task1

Method	OCTIP	F1 score	Rk correlation	Specificity
Early Fusion	✗	0.8037	0.5790	0.8857
Early Fusion	✓	0.8181	0.6094	0.8894
Late Fusion	✗	0.8130	0.6070	0.8967
Late Fusion	✓	0.8223	0.6270	0.9013

Table 3. Reconstruction Accuracy (MSE) on Task 1 Validation Set

Method	Preprocess (OCTIP)	MSE
MAE	✗	0.5572
PPMAE	✗	0.3269
PPMAE	✓	0.1724

(In MAE, we follow the traditional training approach by masking parts of the image and attempting to reconstruct the masked patches. The goal remains to reconstruct the image at time t_1.)

Table 3 shows the image t_1 reconstruction performance, the PPMAE model achieved a MSE of 0.3269 without preprocessing and 0.1724 when preprocessing was applied. In comparison, the traditional MAE resulted in a higher MSE of 0.5572, indicating that the PPMAE model significantly improves reconstruction accuracy, especially when preprocessing is applied.

The reconstruction results of the three categories are shown in Fig. 5. As shown in the figure, the model is still able to produce high-quality predictions and maintain good structural similarity with the original images.

Table 4 shows that the PPMAE-based method performs better across multiple metrics. This indicates that a better reconstruction significantly enhances performance. The approach using PPMAE with finetuning achieves the best overall performance, with a F1 score of 0.6933 and a mean metric value of 0.3748. Meanwhile, the approach using PPMAE without finetuning outperforms the other methods in terms of Rk correlation and weighted quadratic kappa. Additionally, in MAE, the reconstruction of t_1 often fails, with the reconstructed image resembling t_0 more than t_1, which results in most predicted classes being stable.

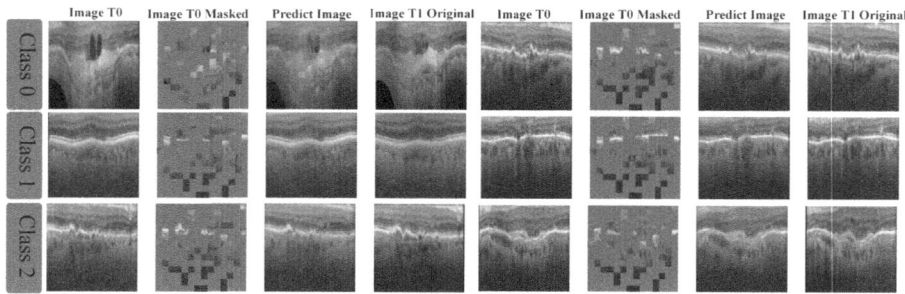

Fig. 5. Visual Comparison of OCT Reconstruction Results by PPMAE

Table 4. Preliminary results on MARIO Challenge - Task2

Method	F1 score	Rk correlation	Specificity	WQK	Mean metrics
MAE + Task 1 models	0.6465	0.0790	0.6936	0.0060	0.3563
PPMAE + Task 1 models	0.6491	0.0938	0.6963	0.0410	0.3701
PPMAE + Task 1 models finetuned on data Task 2	0.6933	0.0923	0.6910	0.0227	0.3748

(In MAE, we follow the traditional training approach by masking parts of the image and attempting to reconstruct the masked patches. The goal remains to reconstruct the image at time t_1.)

4 Discussion and Conclusions

In this work, we presented our solution for the two tasks in the MARIO Challenge. For Task 1, we utilized Late Fusion CNN Network with ResNet50 as the encoder, combined with OCTIP, to achieve the best results in feature extraction and classification accuracy. Leveraging model ensembling further enhanced our performance metrics. For Task 2, we introduced the PPMAE, which predicts future OCT images based on a single exam, demonstrating strong reconstruction and classification performance. Notably, OCTIP played a key role in both tasks by improving image quality, enhancing classification accuracy, and enabling more precise and efficient feature extraction, which also contributed to better image reconstruction. Our methods yielded excellent results, placing us in the Top 10 for both tasks. These outcomes highlight the potential of advanced neural network architectures in developing more personalized and effective treatment strategies for AMD.

However, predicting disease progression from a single exam remains challenging. Adding patient data, such as age, sex, visit number, and eye laterality, did not improve performance. This aligns with our observation that these factors are poorly correlated with disease progression labels. Further research into image reconstruction methods, particularly for generating better future images, is essential to continue improving performance in this field.

Disclosure of Interests. The authors have no competing interests.

References

1. Ashir, A.M., Ibrahim, S., Abdulghani, M., Ibrahim, A.A., Anwar, M.S.: Diabetic retinopathy detection using local extrema quantized haralick features with long short-term memory network. Int. J. Biomed. Imaging **2021**(1), 6618666 (2021)
2. Bhuiyan, A., Wong, T.Y., Ting, D., Govindaiah, A., Souied, E.H., Smith, R.T.: Artificial intelligence to stratify severity of age-related macular degeneration (AMD) and predict risk of progression to late AMD. Transl. Vis. Sci. Technol. **9**(2), 25–25 (2020)
3. Dai, L., et al.: A deep learning system for detecting diabetic retinopathy across the disease spectrum. Nat. Commun. **12**(1), 3242 (2021)
4. Dosovitskiy, A.: An image is worth 16x16 words: transformers for image recognition at scale. arXiv preprint (2020). arXiv:2010.11929
5. Freund, K.B., et al.: Treat-and-extend regimens with anti-VEGF agents in retinal diseases: a literature review and consensus recommendations. Retina **35**(8), 1489–1506 (2015)
6. He, K., Chen, X., Xie, S., Li, Y., Dollár, P., Girshick, R.: Masked autoencoders are scalable vision learners. In: Proceedings of the IEEE/CVF Conference on Computer Vision and Pattern Recognition, pp. 16000–16009 (2022)
7. Hochreiter, S.: Long short-term memory. Neural Comput. **9**(8), 1735–1780 (1997)
8. Holste, G., et al.: Harnessing the power of longitudinal medical imaging for eye disease prognosis using transformer-based sequence modeling. NPJ Digit. Med. **7**(1), 216 (2024)
9. Jonas, J.B., Cheung, C., Panda-Jonas, S.: Updates on the epidemiology of age-related macular degeneration. Asia-Pac. J. Ophthalmol. **6**(6), 493–497 (2017)
10. Lains, I., et al.: Retinal applications of swept source optical coherence tomography (OCT) and optical coherence tomography angiography (OCTA). Prog. Retin. Eye Res. **84**, 100951 (2021)
11. Li, E., Donati, S., Lindsley, K.B., Krzystolik, M.G., Virgili, G.: Treatment regimens for administration of anti-vascular endothelial growth factor agents for neovascular age-related macular degeneration. Cochrane Database Syst. Rev. (5) (2020)
12. Li, Y., et al.: A review of deep learning-based information fusion techniques for multimodal medical image classification. Comput. Biol. Med. 108635 (2024)
13. Rasmussen, A., Sander, B.: Long-term longitudinal study of patients treated with ranibizumab for neovascular age-related macular degeneration. Curr. Opin. Ophthalmol. **25**(3), 158–163 (2014)
14. Rosenfeld, P.J., et al.: Ranibizumab for neovascular age-related macular degeneration. N. Engl. J. Med. **355**(14), 1419–1431 (2006)
15. Thomas, M.: Reconnaissance par apprentissage profond de l'évolution de l'activité néovasculaire des DMLA en tomographie par cohérence optique. Ph.D. thesis, Université de Bretagne Occidentale, Brest (2021)
16. Zhang, P., et al.: Detection and classification of glaucoma in the justraigs challenge: achievements in binary and multilabel classification. In: 2024 IEEE International Symposium on Biomedical Imaging (ISBI), pp. 1–4. IEEE (2024)

Automatic Detection and Prediction of nAMD Activity Change in Retinal OCT Using Siamese Networks and Wasserstein Distance for Ordinality

Taha Emre[1](✉), Teresa Araújo[2], Marzieh Oghbaie[2], Dmitrii Lachinov[2], Guilherme Aresta[2], and Hrvoje Bogunović[2]

[1] Department of Ophthalmology and Optometry, Medical University of Vienna, Vienna, Austria
taha.emre@meduniwien.ac.at

[2] Christian Doppler Laboratory for Artificial Intelligence in Retina, Institute of Artificial Intelligence, Center for Medical Data Science, Medical University of Vienna, Vienna, Austria

Abstract. Neovascular age-related macular degeneration (nAMD) is a leading cause of vision loss among older adults, where disease activity detection and progression prediction are critical for nAMD management in terms of timely drug administration and improving patient outcomes. Recent advancements in deep learning offer a promising solution for predicting changes in AMD from optical coherence tomography (OCT) retinal volumes. In this work, we proposed deep learning models for the two tasks of the public MARIO Challenge at MICCAI 2024, designed to detect and forecast changes in nAMD severity with longitudinal retinal OCT. For the first task, we employ a Vision Transformer (ViT) based Siamese Network to detect changes in AMD severity by comparing scan embeddings of a patient from different time points. To train a model to forecast the change after 3 months, we exploit, for the first time, an Earth Mover (Wasserstein) Distance-based loss to harness the ordinal relation within the severity change classes. Both models ranked high on the preliminary leaderboard, demonstrating that their predictive capabilities could facilitate nAMD treatment management (https://github.com/EmreTaha/Siamese-EMD-for-AMD-Change).

Keywords: Longitudinal change detection · Age-related macular edema · Siamese networks · Ordinal classification

1 Introduction

Neovascular age-related macular degeneration (nAMD) is a progressive exudative disease characterized by the accumulation of fluid in the macula, which can significantly impair vision function [6]. Anti-VEGF treatments have shown

---

T. Emre, T. Araújo and M. Oghbaie—Equal contribution.

great efficacy in mitigating AMD progression, and the positive effects are optimized by reducing the time from fluid onset to treatment, and thus regular follow-up is crucial for successful patient outcomes. The presence and extent of exudative signs, such as intraretinal and subretinal fluid, which are best visible on optical coherence tomography (OCT) images, are relevant markers for anti-VEFG administration [19]. Thus, predicting and accurately detecting changes in neovascularization activity are pivotal tasks for treatment management. The "Monitoring Age-related Macular Degeneration Progression In Optical Coherence Tomography" (MARIO) Challenge, organized as part of MICCAI 2024 [1], aims the development of automatic methods for nAMD progression assessment and evolution prediction.

1.1 Longitudinal Change Detection in Retinal OCT

The manual assessment of changes between image pairs is time-consuming and challenging, highlighting the need for automated systems to detect meaningful changes and reduce specialists' workload. Although machine learning, particularly deep learning, has advanced in automating OCT disease diagnostics, there has been little focus on detecting changes between sequential OCT images.

In medical image analysis, current deep learning methods for change detection usually fall into two categories: 1) Siamese networks [12,13,21], or 2) Graph-based methods [9,18]. In Li *et al.* [13], change detection is approached as a metric-learning problem, using a Siamese neural network to assess changes between two patient visits. A contrastive loss function is applied between features of two images, labeled as *change* or *no change*, to output a pairwise distance between images from two time points. This method was tested on retinal photographs (diabetic retinopathy of prematurity) and knee x-rays (osteoarthritis). In Li *et al.* [12], the same authors used a similar approach to track COVID-19 pulmonary disease severity in chest x-rays.

In contrast, Karward *et al.* [9] proposed a graph-based, anatomy-aware model for tracking changes between chest X-rays by using both local and global anatomical information. This model provides localized comparisons between sequential X-ray images and outperformed Siamese-based models. However, this type of approach is more suitable when the structures are within a rigid region, and thus have a stable anatomical location (e.g., chest x-rays), and are not that suitable for retinal OCT, where pathologies can cause significant tissue deformations.

1.2 Prediction of nAMD Evolution Within 3 Months on OCT

nAMD severity change prediction is crucial for patient follow-up scheduling and timely drug (anti-VEGF) administration. In this aspect, it is essential to have a predictive model that can assess the change in AMD disease activity within a meaningful time-frame from an OCT scan acquired at a visit. Prior work has largely focused on a related task of intermediate AMD progression prediction [2,4,17,22]. The main challenge of training a predictive model of disease progression and treatment response is the training data. Since, the nature of the task is temporal, it is crucial to have follow-up visits from a patient to create a longitudinal dataset. In [2,4,17], they used longitudinal self-supervised learning methods to learn the temporal relations. Additionally, most of the available OCTs have no observable change within short time windows, resulting in severe class imbalance, requiring specialized deep learning loss terms to address this [14].

1.3 Contributions

We contribute to the state of the art in longitudinal retinal OCT assessment in two ways. First, we propose a Siamese-based approach, SiamRETFound, relying on an OCT foundation model (RETFound [23]) and additional pretraining for learning to estimate clinically-relevant changes between two patient B-scans. Second, we propose a classifier model that predicts the nAMD evolution within 3 months. Specifically, our novel loss has 2 parts: a focal loss to address the severe class imbalance, and an Earth Mover's Distance-based (EMD) loss to harness the ordinal relation within the severity change classes. To our knowledge, this is the first study to use EMD for predicting disease severity evolution.

1.4 MARIO Challenge

We assess our methods on the two tasks of the MARIO challenge. Task 1 (T1) aims at developing algorithms to classify changes in disease activity from pairs of retinal OCT B-scans from two visits of patients with nAMD undergoing anti-VEGF treatment, to support decision-making. The classification categories were *Reduced, Stable, Worsened*, or *Other* (Uninterpretable). Based on our qualitative assessment of the provided training set, *Other* seemed to be associated with scenarios that prevented proper clinical assessment, such as the presence of noise, obscured regions, vertical flipping of the scan, poor alignment between image pairs, etc. (Fig. 2(e)–(g)). However, no objective criteria was provided, and there were a few nuanced cases (Fig. 2(g)).

In Task 2 (T2), the goal is to train a predictive model that can accurately predict the change in disease activity within 3 months, given a single B-scan. The prediction categories are *Reduced, Stable* and *Worsened*. Even though the labels and the prediction are on B-scan level, labels are consistent within a volume.

2 Method

2.1 T1: SiamRETFound for Longitudinal Change Detection

We propose SiamRETFound, a Siamese neural network with shared weights for evaluating change between retinal B-scans from two visits of the same patient (Fig. 1). SiamRETFound uses a late fusion approach by extracting and concatenating two feature representations, one for each B-scan. The resulting feature vector, which integrates meaningful features from both time points, is then processed by a classification head to obtain the change detection. Ultimately,

SiamRETFound learns to assess which differences between the extracted features are relevant to identify clinically meaningful changes.

The backbone for feature extraction consists of a pair encoders of the RETFound model, a foundation model pre-trained on retinal OCTs using a masked autoencoder strategy [23]. The encoder is a large vision Transformer (ViT) with 24 Transformer blocks and embedding vector size of 1024. The classification head has a fully connected (FC) layer with 256 neurons and a ReLU activation, 25% dropout, and a FC with 4 output neurons (number of classes).

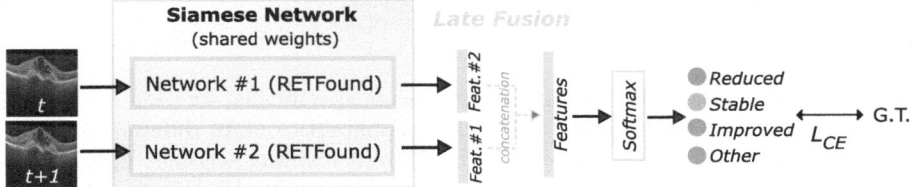

Fig. 1. SiamRETFound approach for longitudinal change detection in retinal OCT.

The full siamese network is first pretrained on a public OCT dataset using a simulated binary change detection task: *change/no change*. The network is then finetuned on the MARIO training data for classification of the target classes.

Datasets and Evaluation. For training and evaluating our model we used the MARIO challenge task 1 development data. This is composed of a training set for which ground truth labels are provided and a validation set for which labels are kept hidden and that is used for ranking participants. The training set 14496 has B-scan pairs from 68 patients, with up to 10 visits per patient. The validation set has 7010 pairs from 34 patients. All OCT volumes were acquired with Spectralis device and longitudinal 3D volumes were anatomically aligned with the follow-up mode from the device.

The public Kermany dataset [10] used for pretraining consists of 100,000 images from patients belonging to one of the following classes: healthy, intermediate AMD, nAMD, and diabetic macular edema (DME). Surrogate binary labels *change/no change* were inferred from the disease classes: if both images have the same disease, the pair was attributed to *no change*, otherwise to *change*.

The evaluation metrics reported in this work include the official MARIO challenge metrics: F1-score (micro-averaging), Specificity and Rk-correlation coefficient (i.e., Mathews correlation coeff. for multiclass setting), along with Balanced accuracy and Cohen's Kappa Coefficient (k) that we added.

Training Details. The MARIO challenge training set was split patient-wise in 90% for training and 10% for validation. To generate the pretraining data from the Kermany dataset, random pairs are taken from the original dataset, for a total of 300,000 image pairs. Both pretraining and finetuning follow the same scheme. Training-time image augmentation is applied, ensuring the same augmentation on both images of a pair. Augmentations include resize-crop (minimum of 85% image size), rotation (up to 15°) and horizontal flipping. Images are resized to 224×224 pixels and the intensities are normalized by the ImageNet mean and standard deviation. Batches are balanced class-wise during training to compensate for the class imbalance. A warm-up is done in the beginning of the training with classification neurons frozen for 50 epochs. The Adam optimizer with an initial learning rate (LR) of 0.001 with an exponential decay was used, and the training loss was the cross-entropy loss. SiamRETFound was finetuned for up to 100 epochs with early stopping based on the average of the challenge evaluation metrics on the held out validation set.

Ensembling. For the MARIO challenge ranking specifically, we used an ensembling strategy by combining the prediction probabilities of 10 different models: 1) 5 SiamRETFound-based models (with different training settings: optimizers, LR, augmentation schemes); 2) 5 Shifted WINdows (SWIN) [15] transformer-based models (5 folds). To obtain a prediction for a test image we took the mean of all models' predictions per class, and chose the class with the maximum-score.

In our SWIN-based models, we first trained a SWIN transformer sequentially on three publicly available OCT datasets: OCTID [5], OCTDL [11], and Kermany [10]. Second, we finetuned the model for multi-label biomarker detection using the OLIVES dataset [16][1]. This method ensures that the model is exposed to a broad spectrum of diseases, enhancing its ability to detect critical biomarkers in B-scans, crucial for change detection.

2.2 T2: WAsserstein Distance for RetInal Disease Ordinality Modeling (WARIO)

Our solution to T2 has 3 main components: masked-autoencoder based pretraining, finetuning with a novel approach to ordinal classes, and postprocessing to ensure volume level consistency based on individual B-scan level predictions.

Pretraining. Pretraining is an important step of the current deep learning pipeline. In AMD disease progression tasks, numerous works showed the benefit of pretraining [2,4]. Recently, masked auto-encoders [7] (MAE) achieved the state-of-the-art in vision tasks. Following this trend, we pretrain our models with MAE by masking 75% of B-scans (Fig. 6) from a combined dataset of T1 and T2.

Finetuning. Addressing T2 effectively requires accounting for (i) the class imbalance due to large number of *Stable*, and (ii) the ordinal relationship between the classes (*Reduced-Stable-Worsened*). While the class imbalance poses a challenge, the ordinality provides an opportunity for additional training signal. For the class imbalance, we adapted Focal Loss [14], which dynamically calculates weights for hard and easy examples, so that hard to classify samples will contribute more:

---

[1] Same splits as the IEEE SPS VIP Cup 2023: Ophthalmic Biomarker Detection.

$$\ell_{\text{focal}} = -\alpha \sum_{i=0}^{C-1} y_i(1-\hat{p}_i)^\gamma \log(\hat{p}_i), \tag{1}$$

where C is the number of classes, and γ is the focusing hyper-parameter for hard to classify samples. Additionally we undersample the majority class (*Stable*) during finetuning.

For the ordinality we use discrete earth-mover's distance [8,20] (EMD, also known as Wasserstein-distance). Unlike the cross-entropy loss, it takes inter-class relations (e.g. ranking) into account when calculating the loss. In T2, the classes are ordered by their definition, as *Reduced-Stable-Worsened*. EMD calculates cumulative mass need to be moved to make one distribution similar to another:

$$\ell_{\text{EMD}} = \left(\frac{1}{C} \sum_{i=0}^{C-1} |\text{CDF}_y(i) - \text{CDF}_{\hat{p}}(i)|^2 \right)^{1/2}, \tag{2}$$

where $\text{CDF}_p()$ is the cumulative distribution function calculated from the class probabilities or one-hot encoded target vector. The final training loss is the combination of these two loss terms with equal weighting.

Postprocessing (Ensembling). After pretraining and finetuning on 3-fold splits, the predictions are ensembled as follows. We make a single prediction from different training split folds in a B-scan level, and then an eye level prediction from available B-scans of a particular OCT volume. In general, the trained networks tend to predict *Stable* due to the imbalance. In order to alleviate this, the final prediction from the 3 outputs is set to class *Stable* if only all 3 models predict *Stable*, otherwise it is set to the majority vote.

In T2, the labels are provided at the OCT volume level, i.e. B-scans from a particular visit scan of a patient have the same label. To ensure this consistency in our predictions, we apply a majority voting. In order to avoid only predicting the majority *Stable* class, a volume level label is set to *Stable* if at least 80% (the class ratio) of the B-scan level predictions are *Stable*. If not, volume level label is the majority prediction. Then we set all B-scans of that volume to a single label, and reported the predictions in this manner.

Datasets and Evaluation. T2 training data consisted of 8082 B-scans from 330 volumes for 61 patients. The provided validation split contained 3822 B-scans from 163 volumes for 29 patients. All OCT volumes are Spectralis and consecutive 3D volumes are registered. For metrics, the official MARIO challenge metrics: F1-score (micro-averaging), Specificity, Quadratic-weighted Kappa and Rk-correlation coefficient, and additional Balanced accuracy are used.

Training Details. We use ViT [3] base model with a single FC prediction head. It has 12 MHSA blocks with 12 parallel heads, embedding size of 768 and 16×16 patch size. The prediction head uses global average pooled embedding

Table 1. Results of our methods in the MARIO challenge validation sets for T1 and T2. For T1, Kappa is Cohen's Kappa; for T2, it is Quadratic-weighted Kappa.

Metric	Task 1 (T1)		Task 2 (T2)	
	SiamRETFound	Ensemble	WARIO	Ensemble
Balanced Acc.	0.713	0.691	0.336	0.485
Kappa^{T2}	0.642	0.656	0.133	0.223
F1 Score^{T1,T2}	0.817	0.833	0.720	0.705
Rk-Correlation^{T1,T2}	0.642	0.657	0.004	0.206
Specificity^{T1,T2}	0.917	0.911	0.666	0.722
Average$^{(*)}$	0.792	0.801	0.351	0.469

features instead of the CLS token. We first run MAE pretraining for 800 epochs with a decayed LR of 3E-4, weight decay of 1E-2, and batch size of 256. We use AdamW as the optimizer with its Beta2 hyperparameter set to 0.95. We then finetune end-to-end for 200 epochs with a warm-up of 20 epochs, a LR of 2.5E-4 and a weight decay of 1E-4. We used AdamW as the optimizer with Beta2 of 0.999. We only use 2D B-scans and considered each B-scan i.i.d (even from the same patient) during training. Training augmentations include translation, small rotation, and horizontal flipping. Each image is resized to 224 × 224 pixels.

3 Results and Discussion

3.1 T1: Longitudinal Change Detection

We showed that SiamRETFound is able to correctly classify longitudinal change for the majority of the OCT pairs (Table 1). The model is able to capture very subtle changes (Fig. 2(a), (b)). Specifically for the change classes (*Reduced*, *Stable* and *Worsened*), which can be interpreted as an ordinal classification problem, we verified most errors occur between neighboring classes and little *Reduced/Worsened* cases are confounded (Appendix - Fig. 3). For the class Other, the majority of the confusion occurs with *Stable*, which can be attributed to the not so clear definition of *Other* (e.g., how much noise or obscured area constitutes an uninterpretable pair) (Fig. 2(g)).

The occlusion sensitivity maps (Appendix - Fig. 5) suggest that the SiamRETFound model is capturing relevant features to classify change between image pairs, mostly focusing on fluid regions.

Despite SiamRETFound approach being better in terms of overall classification performance (inferred by the confusion matrix and the balanced accuracy and k metrics), concerning the MARIO challenge metrics the ensemble version outperformed the single model prediction. On the challenge leaderboard in the development phase we ranked 4th considering the mean of the evaluation metrics, with a score of 0.801 (scores of the top 3 teams: 0.828/0.805/0.804).

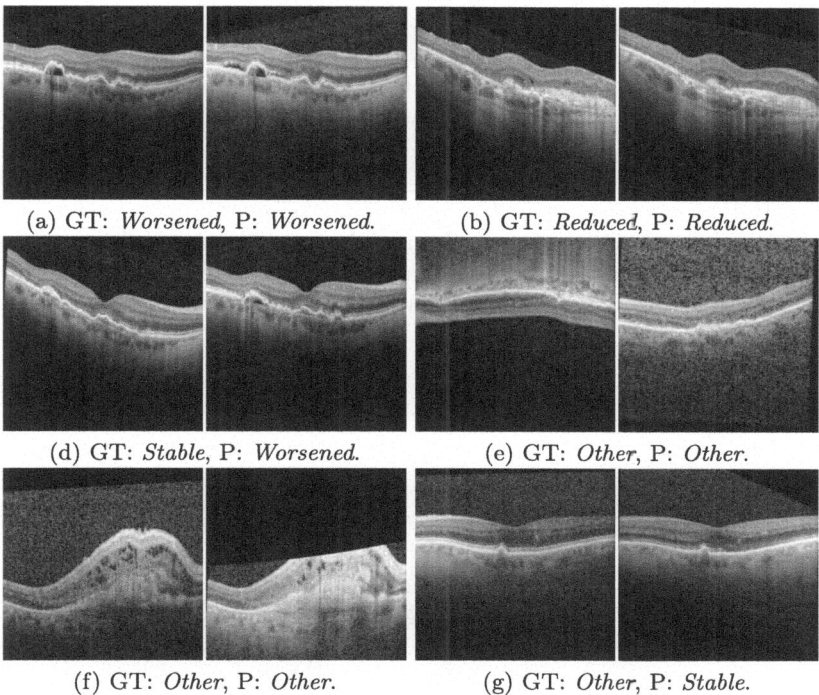

Fig. 2. Examples of SiamRETFound predictions (GT: ground truth, P: prediction).

3.2 T2: Prediction of AMD Evolution Within 3 Months

In Task 2, the goal is to detect the change in AMD severity in 3 months from a provided past visit scan. We found that, it is crucial to have a strong pretraining step and correct loss terms. Only the combination of focal loss and EMD loss (WARIO) prevent the network from only predicting the *Stable* class. In Table 1, it is clear that WARIO is still heavily biased towards the majority class, highlighting the importance of a postprocessing step. Clinically we know that AMD change is a OCT volume level assessment from which we concluded that the B-scan level predictions need to be consistent along an OCT volume. We enforced this consistency in the postprocessing step combined with an ensemble of 3-fold data split. Even though F1-score dropped slightly, WARIO improved along other metrics. It is important to highlight that postprocessing improved Rk-correlation the most which is generally used in imbalanced classification problems. At the end, WARIO ranked 2nd in the challenge learderboard.

4 Conclusions

MICCAI'24 MARIO challenge aims to model AMD severity change in 2 setups: (i) predicting the change by comparing B-scans from two time points, (ii) predicting the change after 3 months from a single B-scan. We proposed SiamRET-Found, a siamese-based approach that was able to effectively classify longitudinal change in OCT pairs, even for very subtle cases. The proposed model captured relevant features to classify change between B-scan pairs, mostly focusing on exudative regions. Predicting the change in OCTs from a single past visit is extremely challenging task. Our method, WARIO, uses focal loss for class imbalance and EMD loss to exploit ordinal relation in the severity change classes, preceded by a strong MAE based pretraining. The proposed methods show potential to facilitate the clinical workflow on nAMD diagnosis from retinal OCTs and allow timely and thus more effective treatment planning (Fig. 4).

A Appendix

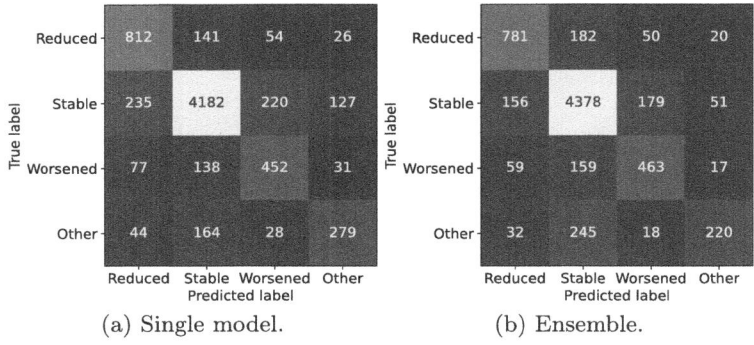

Fig. 3. Confusion matrices for Longitudinal change detection (MARIO Task 1).

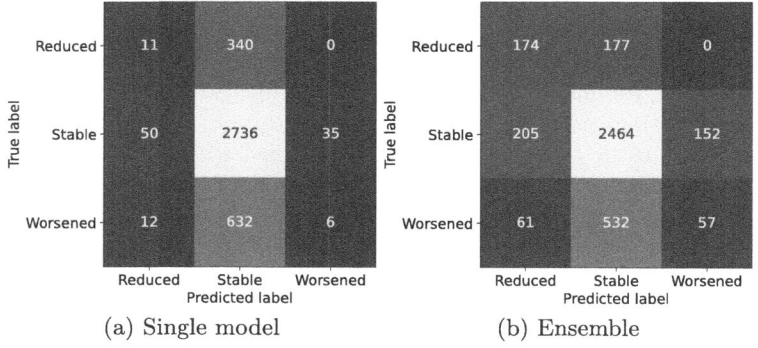

Fig. 4. Confusion matrices for AMD evolution prediction (MARIO Task 2)

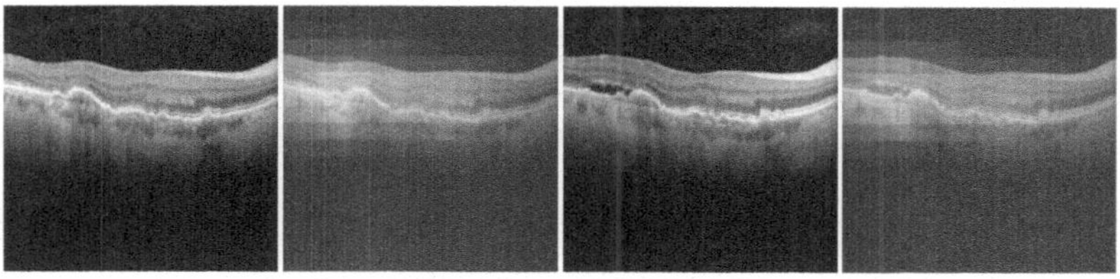

(a) GT: *Worsened*, P: *Worsened*.

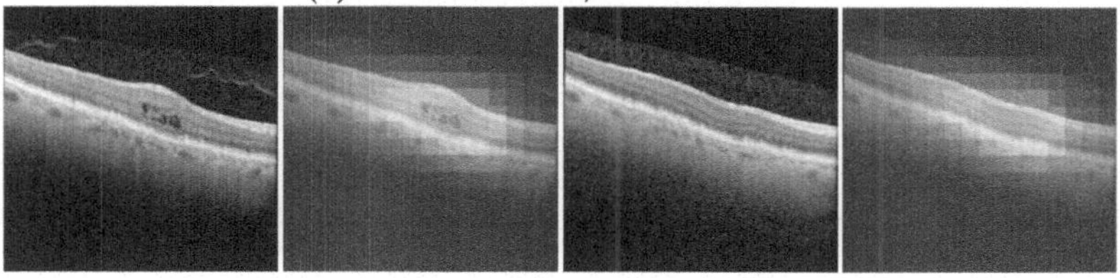

(b) GT: *Reduced*,, P: *Reduced*.

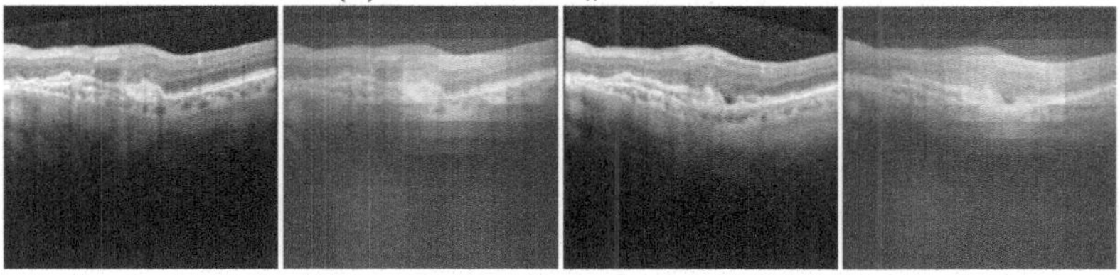

(c) GT: *Worsened*, P: *Worsened*.

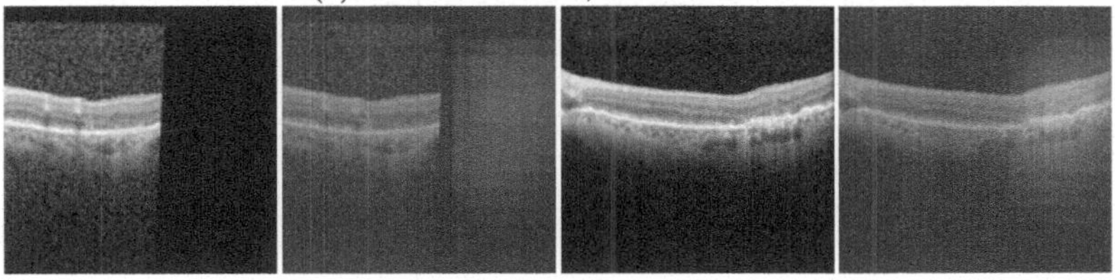

(d) GT: *Other*, P: *Other*.

Fig. 5. Occlusion map sensitivity. Map values are from 0 (blue) to 1 (red). Lower values indicate the occluded region impacted more the final prediction. (Color figure online)

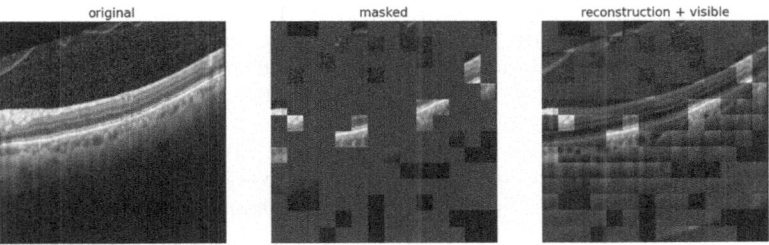

Fig. 6. MAE pretraining with masking and reconstruction.

References

1. MARIO challenge (2024). htttps://www.codabench.org/competitions/2852
2. Chakravarty, A., et al.: Morph-SSL: self-supervision with longitudinal morphing for forecasting AMD progression from OCT volumes. IEEE Trans. Med. Imaging (2024)
3. Dosovitskiy, A., et al.: An image is worth 16x16 words: transformers for image recognition at scale. In: International Conference on Learning Representations (2021)
4. Emre, T., Chakravarty, A., Rivail, A., Riedl, S., Schmidt-Erfurth, U., Bogunović, H.: Tinc: temporally informed non-contrastive learning for disease progression modeling in retinal oct volumes. In: International Conference on Medical Image Computing and Computer-Assisted Intervention, pp. 625–634. Springer (2022)
5. Gholami, P., Roy, P., Parthasarathy, M.K., Lakshminarayanan, V.: OCTID: optical coherence tomography image database. Comput. Electr. Eng. **81**, 106532 (2020)
6. Guymer, R.H., Campbell, T.G.: Age-related macular degeneration. Lancet **401**(10386), 1459–1472 (2023)
7. He, K., Chen, X., Xie, S., Li, Y., Dollár, P., Girshick, R.: Masked autoencoders are scalable vision learners. In: Proceedings of the IEEE/CVF Conference on Computer Vision and Pattern Recognition, pp. 16000–16009 (2022)
8. Hou, L., Yu, C.P., Samaras, D.: Squared earth mover's distance-based loss for training deep neural networks. arXiv preprint (2016). arXiv:1611.05916
9. Karwande, G., Mbakwe, A.B., Wu, J.T., Celi, L.A., Moradi, M., Lourentzou, I.: CheXRelNet: an anatomy-aware model for tracking longitudinal relationships between chest x-rays. In: Wang, L., Dou, Q., Fletcher, P.T., Speidel, S., Li, S. (eds.) MICCAI 2022. LNCS, vol. 13431, pp. 581–591. Springer, Cham (2022). https://doi.org/10.1007/978-3-031-16431-6_55
10. Kermany, D.S., Goldbaum, M., Cai, W.: Identifying medical diagnoses and treatable diseases by image-based deep learning. Cell **172**(5), 1122–1131 (2018). https://doi.org/10.1016/j.cell.2018.02.010
11. Kulyabin, M., et al.: Octdl: optical coherence tomography dataset for image-based deep learning methods. Sci. Data **11**(1), 365 (2024)
12. Li, M.D., et al.: Automated assessment and tracking of COVID-19 pulmonary disease severity on chest radiographs using convolutional siamese neural networks. Radiol. Artif. Intell. **2**(4), 1–39 (2020). https://doi.org/10.1148/ryai.2020200079
13. Li, M.D., et al.: Siamese neural networks for continuous disease severity evaluation and change detection in medical imaging. NPJ Digit. Med. **3**(1), 1–9 (2020). https://doi.org/10.1038/s41746-020-0255-1
14. Lin, T.Y., Goyal, P., Girshick, R., He, K., Dollár, P.: Focal loss for dense object detection. In: Proceedings of the IEEE International Conference on Computer Vision, pp. 2980–2988 (2017)
15. Liu, Z., et al.: Swin transformer: Hierarchical vision transformer using shifted windows. In: Proceedings of the IEEE/CVF International Conference on Computer Vision, pp. 10012–10022 (2021)
16. Prabhushankar, M., Kokilepersaud, K., Logan, Y.Y., Trejo Corona, S., AlRegib, G., Wykoff, C.: Olives dataset: ophthalmic labels for investigating visual eye semantics. Adv. Neural. Inf. Process. Syst. **35**, 9201–9216 (2022)
17. Rivail, A., et al.: Modeling disease progression in retinal OCTS with longitudinal self-supervised learning. In: Predictive Intelligence in Medicine: Second International Workshop, PRIME 2019, Held in Conjunction with MICCAI 2019, Shenzhen, China, 13 October 2019, Proceedings 2, pp. 44–52. Springer (2019)

18. Rochman, S., Szeskin, A., Lederman, R., Sosna, J., Joskowicz, L.: Graph-based automatic detection and classification of lesion changes in pairs of CT studies for oncology follow-up. Int. J. Comput. Assist. Radiol. Surg. (2023). https://doi.org/10.1007/s11548-023-03000-2
19. Schmidt-Erfurth, U., Waldstein, S.M.: A paradigm shift in imaging biomarkers in neovascular age-related macular degeneration. Prog. Retin. Eye Res. **50**, 1–24 (2016)
20. Talebi, H., Milanfar, P.: NIMA: neural image assessment. IEEE Trans. Image Process. **27**(8), 3998–4011 (2018)
21. To, M.-S., Sarno, I.G., Chong, C., Jenkinson, M., Carneiro, G.: Self-supervised lesion change detection and localisation in longitudinal multiple sclerosis brain imaging. In: de Bruijne, M., et al. (eds.) MICCAI 2021. LNCS, vol. 12907, pp. 670–680. Springer, Cham (2021). https://doi.org/10.1007/978-3-030-87234-2_63
22. Yim, J., et al.: Predicting conversion to wet age-related macular degeneration using deep learning. Nat. Med. **26**(6), 892–899 (2020)
23. Zhou, Y., Chia, M.A., Wagner, S.K.: A foundation model for generalizable disease detection from retinal images. Nature **622**(7981), 156–163 (2023). https://doi.org/10.1038/s41586-023-06555-x

Exploring the Use of Off-the-Shelf AI Models for Complex Medical Tasks: ResNet18 for Predicting Age-Related Macular Degeneration

Amerens A. Bekkers[1], Nina M. van Liebergen[1], and Hugo J. Kuijf[1,2(✉)]

[1] TNO, The Hague, The Netherlands
hugo.kuijf@tno.nl
[2] Image Sciences Institute, UMC Utrecht, Utrecht, The Netherlands

Abstract. Age-related macular degeneration (AMD) is a progressive eye disease that affects close to 200 million people worldwide. Treatment that works against the Vascular Endothelial Growth Factor (VEGF) can slow disease progression and even improve visual function. AI has not yet been exhaustively explored as a tool to reliably assess evolution in neovascular activity, needed to correctly implement anti-VEGF treatment strategies. In this work, the off-the-shelf ResNet18 model is evaluated on the complex task of classification and prediction of AMD as a part of the MICCAI 2024 MARIO challenge. In a first exploration, a baseline ResNet18 obtained an accuracy of 80% and 87% for the two different tasks of the challenge. Validation results were provided by the challenge organizers and shows a somewhat lower performance, but sufficient to be selected amongst the top-ranking finalists. The baseline models for both tasks was then retrained with a 5-fold cross-validation on both the training and validation data for the final phase of the challenge, which again showed promising performances of an average accuracy of 84.6% and 90.7% for classification and prediction, respectively. In conclusion, an off-the-shelf AI model gives a reasonable performance when applied to a domain-specific task. Even though this provides a quick first impression of the data and the use of AI, it does not match the performance of models that are tailored to this specific task.

Keywords: Artificial Intelligence · Age-related Macular Degeneration · ResNet

1 Introduction

Age-related macular degeneration (AMD) is a common medical condition that affects close to 200 million people worldwide [1,2]. Both genetic and environmental risk factors are at play, making it a complex multifactorial pathology. It is a progressive degeneration of the central part of the retina (the macula).

Close to 20% of the patients suffer from advanced stages of AMD (atrophy and neovascularistion), leading to severe blurred vision and even blindness [2]. Antivascular endothelial growth factor (anti-VEGF) treatments have proven their ability to slow disease progression and even improve visual function in neovascular forms of AMD [3,4]. Minimizing the time between diagnosis and start of treatment optimizes the effectiveness of treatment, as well as regular check-ups and additional treatments when necessary [5]. Reliable detection of evolution in neovascular activity [6] by monitoring exudative signs is crucial for the correct implementation of anti-VEGF treatment strategies. Currently, there is limited work on using artificial intelligence (AI) to predict the development of the AMD in close monitoring of the patient undergoing anti-VEGF treatments.

Optical coherence tomography (OCT) is a non-invasive imaging technique to assess AMD. AI-based OCT image analysis could support automated diagnosis and prognosis of AMD. To objectively and quantitatively compare AI technology for the assessment of AMD, [2] initiated the biomedical image analysis challenge MARIO ("Monitoring Age-related Macular Degeneration Progression In Optical Coherence Tomography") [2,7]. Biomedical image analysis challenges provide an objective way to validate and evaluate new algorithms and have been organized for a variety of tasks, ranging from brain disease to full body tasks [8–10]. The MARIO challenge includes a diverse, longitudinal dataset acquired amongst French and Algerian populations. The challenge consists of two tasks:

1. classify the evolution of AMD between two paired OCT images, as either 'reduced', 'stable', 'worsened' (or if the images are 'uninterpretable')
2. predict based on one OCT image if the three-month follow-up will be 'reduced', 'stable', 'worsened'

The aim of this work is to evaluate the performance of an off-the-shelf AI model as a tool to quickly achieve a first impression of its efficacy in the medical domain. A pre-trained ResNet [11,12] model is used to extract features from the OCT images, which are next used for AMD classification and prediction in task 1 and 2.

2 Methods

Task 1 and Task 2 require a tailored approach, because of the different nature of the classification and prediction tasks; and the different data used for each task. Common in both approaches is the use of a pre-trained ResNet for feature extraction, which are weighted differently per task, and used for the final classification or prediction.

The MARIO challenge is divided into two phases: a development phase and the final phase. Participants get access to a training dataset (images and ground truth labels) to develop their method. Next, the validation dataset (images only) is released and had to be processed by the participants. Based on the performance on this validation dataset, the top-ranking methods were invited for the final phase. In the final phase, the ground truth labels of the validation data

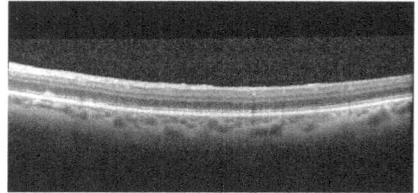
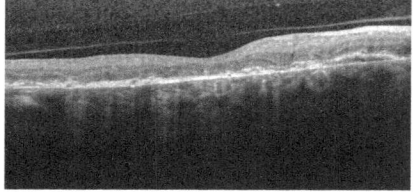

Fig. 1. Example OCT Image 1.   **Fig. 2.** Example OCT Image 2.

were released as well, so participants could re-train and further optimize their methods. At the time of writing this manuscript, the final evaluation has not been performed yet and will be presented later in a separate publication by the challenge organizers.

2.1 Data

A detailed description of the data is provided by [2]. In short, OCT images were acquired at two hospitals in France and Algeria (examples shown in Figs. 1 and 2. Ground truth labels indicating '0 = reduced', '1 = stable', '2 = worsened' disease progression (or if the images are '3 = uninterpretable') were provided. The exact label distribution is not provided, but is approximately 4:20:3:2 for the four labels (based on the French image pairs).

For Task 1, data was provided as image pairs of the same patient with the associated label. Since patients underwent imaging more often, for both eyes, with multiple images per acquisition session; many image pairs were provided (for the French training data: 68 patients resulting in 14.496 image pairs).

For Task 2, only the first image of an image pair was provided (pairs with a three month gap) with the associated label.

For the distribution of classes in each dataset, refer to Table 1.

2.2 Development Phase – Task 1

During the development phase, we explored four different types of models for Task 1: a baseline model and three variations.

Table 1. Class distribution in the provided training data for both Tasks 1 and 2.

Class	Task 1 Distribution	Task 2 Distribution
0	15%	13%
1	65%	81%
2	11%	6%
3	9%	-

Baseline Model: The architecture of the Baseline Model has a Siamese architecture, where two neural networks work in tandem on two different input vectors. In this case, is uses two pre-trained ResNet18 models that process different input data streams. The first model gets as input the images taken on $t = 0$, the second model gets as input all the images from $t = 1$. After training, the outcome features from both models are then combined and fed into a new fully connected layer that makes the final classification.

Extra Layer: To improve the model, we added an extra layer to the baseline model after combining the two ResNet models. We expect that this will improve the model because this extra layer could make connections between the two models and learn from the feature interaction. We decided to add only one extra layer, as adding two extra layers showed no significant improvement after testing.

Balanced Batches: To optimize our model in a different way, we focus at the unequal distribution of classes detailed in Sect. 2.1. To make sure that the model learns about every class equally, we train a model with balanced batches: ensuring that each batch of data used during training contains an equal or proportionate number of samples from each class. Balanced batches are achieved by iterating through separate data loaders for each class. During each iteration, a batch is drawn from each loader, and then the samples are concatenated together to form a single batch that contains an equal number of samples from each class. This approach helps the model to learn equally from all classes, preventing it from becoming biased towards the more frequent 'stable' class.

Balanced Batches and Augmented Data: Some data augmentation was applied to artificially expand the diversity of the training dataset by applying various transformations. Data augmentation was implemented through a series of image transformations, including random horizontal and vertical flips, rotations, brightness/contrast (with PyTorch ColorJitter), and resized cropping. The datasets are then combined to create a comprehensive training set that includes both the original and augmented images. We combine this method with our balanced batches approach. For Task 1, we generate 15,000 samples per class. Thus, we end up with a dataset that is four times as big as the original dataset of which the classes are balanced.

2.3 Development Phase – Task 2

First, the baseline model is a pre-trained ResNet18 model that was trained on 70% of the training data. Second, the same model is retrained with balanced batches, to exclude the effect of the class distribution; similar to Task 1. Lastly, the balanced batches model was extended with augmented data by *torchvision.transforms*; also similar to Task 1.

2.4 Validation by Challenge Organisers

The images from the validation set were released by the challenge organisers, without the ground truth labels. They were process by the methods for Task 1

Table 2. Results of Task 1 for the four models: baseline, with an extra linear layer, with balanced batches, and with balanced batches plus augmentation. (#img) is the number of image pairs.

Class	Basis Model			Extra Layer			Support (#img)
	Precision	Recall	F1-Score	Precision	Recall	F1-Score	
0	0.71	0.69	0.70	0.71	0.71	0.71	652
1	0.84	0.93	0.89	0.88	0.92	0.90	2838
2	0.86	0.41	0.55	0.81	0.50	0.62	462
3	0.69	0.63	0.66	0.65	0.72	0.68	397
Accuracy	0.81			0.83			4349
Macro Avg	0.77	0.66	0.70	0.76	0.71	0.73	4349
Weighted Avg	0.81	0.81	0.80	0.83	0.83	0.82	4349
Class	Balanced Model			Balanced Augmented			Support (#img)
	Precision	Recall	F1-Score	Precision	Recall	F1-Score	
0	0.79	0.66	0.72	0.79	0.77	0.78	648
1	0.85	0.94	0.89	0.94	0.87	0.90	2825
2	0.79	0.61	0.69	0.63	0.86	0.73	489
3	0.78	0.56	0.65	0.73	0.83	0.77	389
Accuracy	0.83			0.85			4351
Macro Avg	0.80	0.69	0.74	0.77	0.83	0.79	4351
Weighted Avg	0.83	0.83	0.82	0.86	0.85	0.85	4351

and Task 2; and the results were submitted for independent evaluation by the challenge organisers.

2.5 Final Phase

After validation by the challenge organisers, the ground truth labels from the validation data were also released and re-training of the models with this data was encouraged. The baseline model for both Task 1 and Task 2 was re-trained and evaluated with a 5-fold cross-validation strategy on all available data: the training and validation sets. The training and validation data were randomly shuffled before creating the subsets for the cross-validation.

3 Results

3.1 Development Phase

Task 1. The model with both balanced batches and augmentation outperforms others by achieving the highest overall accuracy (0.85) and the best macro and weighted averages for recall and F1-score as shown in Table 2. It improves performance in handling class imbalances, especially enhancing recall and F1-score

Table 3. Results of Task 2 for the three models: baseline, with balanced batches, and with balanced batches plus augmentation. #img is the number of images/

Class	Baseline Model			Balanced Model			Support (#img)
	Precision	Recall	F1-Score	Precision	Recall	F1-Score	
0	0.72	0.56	0.63	0.64	0.80	0.71	320
1	0.89	0.96	0.92	0.95	0.91	0.93	1967
2	0.61	0.30	0.40	0.66	0.64	0.65	138
Accuracy		0.87			0.88		2425
Macro Avg	0.77	0.66	0.70	0.76	0.71	0.73	2425
Weighted Avg	0.81	0.81	0.80	0.83	0.83	0.82	2425

Class	Balanced Augmented Model			Support (#img)
	Precision	Recall	F1-Score	
0	0.64	0.78	0.70	320
1	0.95	0.91	0.93	1967
2	0.66	0.77	0.71	138
Accuracy		0.88		2425
Macro Avg	0.75	0.82	0.78	2425
Weighted Avg	0.90	0.88	0.89	2425

for classes with initially lower metrics, such as Class 2 and Class 3. This performance is as expected, since the balanced batches and data augmentation are equalizing the performance for the different classes.

Task 2. The balanced batches method combined with data augmentation performed best out of the three experiments, with an overall accuracy of 0.88. The effect of the balanced batches resulted in the largest increase in performance as visible in the and smaller inter-class differences in precision, recall and F1-score as compared to the baseline model as shown in Table 3.

3.2 Validation by Challenge Organisers

The results of Tasks 1 and 2 were amongst the top-10 finalists as awarded by the challenge organisers. The performance of the models on the validation data was lower than during the development phase. The baseline model for Task 1 performed best with an F1-score of 0.74979, Rk-correlation score of 0.51055, specificity of 0.88694, and a Quadratic-weighted Kappa of 0.71576. The baseline model for Task 2 performed best with an F1-score of 0.65568, Rk-correlation score of 0.03282, specificity of 0.67487, and a Quadratic-weighted Kappa of 0.03628.

3.3 Final Phase

The average results of Task 1 and Task 2 based on a 5-fold cross validation on the training and validation datasets combined can be found in Table 4.

Table 4. Average Performance Metrics Final Phase

Metric	Task 1	Task 2
Average Test Accuracy (%)	0.85	0.91
Average F1 Score	0.85	0.91
Average Rk-Correlation	0.69	0.73
Average Specificity	0.92	0.90
Average Quadratic-weighted Kappa	0.64	0.71

4 Discussion

In the context of Task 1 and Task 2 of the MARIO challenge, we evaluated the performance of an off-the-shelf ResNet model with modifications for AMD assessment.

For both Task 1 and Task 2, the balanced batches method increases the performance of the models in the development phase. Note that the four models in Task 1 were evaluated on different train/test splits, hence these results can only be used as an indication. The balanced batches models likely perform best because this method removes the bias from the unbalanced datasets as shown in Table 1. However, balancing the batches could also lead to a potential bias towards the classes that are underrepresented in the dataset. The model might predict all classes with equal frequency, which does not accurately reflect the distribution of the real dataset. Interestingly, the baseline model showed best performance on the validation set, but was still rather low. This discrepancy between the results based on the training data and the validation data can probably attributed to overfitting. Whether this is resolved by training on the training and validation set during the final phase will become clear after the final challenge results on the test set are made available by the challenge organisers.

The results show that the model for Task 2 has slightly higher accuracy than the model for Task 1, likely because Task 2 is more complex. The slightly better accuracy for Task 2 could be attributed to differences in model architecture and the possibility that Task 1 experiences more overfitting than Task 2.

While aiming for the highest scores in the validation phase, it is important to keep considering the clinical relevance of the outcomes. It might be more clinically relevant to predict the class referring to AMD progression more accurately than the class that refers to stable or reduced AMD. Most of the patients, as is represented in the dataset, have a stable or reduced AMD. However, patients with progressive AMD require the most attention and therefore the most accurate prediction and classification of the disease status.

In conclusion, the aim of this work was to evaluate the performance of an off-the-shelf AI model as a tool to quickly achieve a first impression of its efficacy in this medical use-case. In this case, the ResNet model gives a reasonable performance, but its performance will likely not match the performance of dedicated models that are tailored to the specific task. Using off-the-shelf models could provide a way to quickly get a first impression of the task, data, and the promise of AI on solving the task.

Link to GitHub repository: https://github.com/ninamalou/TONIC-MICCAI-2024/tree/main

References

1. Jonas, J.B., Cheung, C.M.G., Panda-Jonas, S.: Updates on the epidemiology of age-related macular degeneration. Asia Pac. J. Ophthalmol. (2017). ISSN: 2162-0989. https://doi.org/10.22608/apo.2017251
2. Zeghlache, R., et al.: Monitoring age-related macular degeneration progression in optical coherence tomography (2024). https://doi.org/10.5281/zenodo.10992295
3. Rosenfeld, P.J., et al.: Ranibizumab for neovascular age-related macular degeneration. N. Engl. J. Med. **355**(14), 1419-1431 (2006). ISSN: 1533-4406. https://doi.org/10.1056/nejmoa054481
4. Lim, L.S., et al.: Age-related macular degeneration. Lancet **379**(9827), 1728–1738 (2012). ISSN: 0140-6736. https://doi.org/10.1016/s0140-6736(12)60282-7
5. Freund, K.B., et al.: Treat-and-extend regimens with anti-VEGT agents in retinal diseases: a literature review and consensus recommendations. Retina **35**(8), 1489–1506 (2015). ISSN: 0275-004X. https://doi.org/10.1097/iae.0000000000000627
6. Li, E., et al.: Treatment regimens for administration of anti-vascular endothelial growth factor agents for neovascular age-related macular degeneration. Cochrane Database Syst. Rev. **2020**(5) (2020). ISSN: 1465-1858. https://doi.org/10.1002/14651858.cd012208.pub2
7. Maier-Hein, L., et al.: BIAS: transparent reporting of biomedical image analysis challenges. Med. Image Anal. **66**, 101796 (2020). ISSN: 1361-8415. https://doi.org/10.1016/j.media.2020.101796
8. Timmins, K.M., et al.: Comparing methods of detecting and segmenting unruptured intracranial aneurysms on TOF-MRAS: the ADAM challenge. Neuroimage **238**, 118216 (2021). ISSN: 1053-8119. https://doi.org/10.1016/j.neuroimage.2021.118216
9. Kuijf, H.J., et al.: Standardized assessment of automatic segmentation of white matter hyperintensities and results of the WMH segmentation challenge. IEEE Trans. Med. Imaging **38**(11), 2556–2568 (2019). https://doi.org/10.1109/TMI.2019.2905770
10. Antonelli, M., et al.: The medical segmentation decathlon. Nature commun. **13**(1), 4128 (2022)
11. He, K., et al.: Deep Residual Learning for Image Recognition (2015). arXiv:1512.03385 [cs.CV]
12. He, K., et al.: Delving Deep into Rectifiers: Surpassing Human-Level Performance on ImageNet Classification (2015). arXiv:1502.01852 [cs.CV]

AI-Driven Analysis of Sequential OCT Images for Detecting Neovascular Activity in Age-Related Macular Degeneration to Optimize Anti-VEGF Therapy

Yi Ding

The University of Edinburgh, Edinburgh EH8 9YL, UK
scyydd4@outlook.com

Abstract. Age-related Macular Degeneration (AMD) is a leading cause of severe visual impairment and blindness in developed countries, particularly affecting individuals over the age of 65. In this study, we propose a novel AI-driven approach that leverages sequential optical coherence tomography (OCT) images to detect neovascular activity and monitor disease evolution in AMD patients. By analyzing paired OCT scans, our model is designed to identify subtle changes indicative of disease worsening or improvement, enabling more precise and individualized treatment planning. This approach was rigorously tested in the Monitoring Age-related Macular Degeneration Progression In Optical Coherence Tomography for Task 1, part of the MICCAI 2024 challenge, where it achieved second place in the preliminary round, demonstrating its efficacy and potential in clinical applications. The results underscore the model's ability to enhance the decision-making process in AMD management, paving the way for more effective use of anti-Vascular Endothelial Growth Factor (VEGF) therapy therapy.

Keywords: Age-related macular degeneration · Machine learning · Optical coherence tomography

1 Introduction

Age-related Macular Degeneration (AMD) is a leading cause of vision impairment, affecting approximately 196 million people worldwide, predominantly in those aged 50 and above [5]. The disease, particularly in its advanced forms geographic atrophy and neovascular AMD (nAMD)-is a major cause of severe vision loss in developed countries [8]. The introduction of Anti-Vascular Endothelial Growth Factor (anti-VEGF) therapies has significantly improved the management of nAMD, with early diagnosis and continuous monitoring being crucial for effective treatment [11].

Despite advancements in optical coherence tomography (OCT), a standard imaging technique for detecting exudative signs of nAMD, there remains a need for automated systems to monitor disease progression in patients under treatment [12,13]. Current AI-driven approaches have primarily focused on the initial diagnosis of AMD, with limited attention to the ongoing assessment of patients receiving anti-VEGF therapy [2].

In this study, we propose an innovative AI framework, including a custom-designed Siamese network, **OCT-DiffNet**, taking as input two consecutive OCT images and outputs a classification label, optimized for detecting subtle changes in OCT images over time. This architecture uniquely accelerates convergence by focusing on the differences between consecutive OCT scans, enabling more precise tracking of disease progression. The model adopts advanced CNN architectures like ConvNeXt V2 [15], alongside robust data augmentation techniques and a majority voting ensemble method, to enhance predictive accuracy.

Our approach was validated through its outstanding performance, achieving second place in the preliminary round of the MICCAI MARIO TASK1 challenge. This result underscores the potential of our model in improving the management and treatment planning for AMD, particularly in real-time monitoring scenarios. By advancing the state of AI in ophthalmology, this work offers a significant contribution to the on-going efforts to combat vision loss due to AMD.

2 Related Works

AI Models for AMD Detection. Recent advancements in CNN architectures have significantly improved the analysis of OCT images for AMD detection. ConvNeXt V2 [15], inspired by transformer models, modernizes the ResNet structure by incorporating Layer Normalization and advanced training strategies, positioning it as a strong competitor to transformer-based models like Vision Transformer (ViT) [1] and Swin Transformer[9] while maintaining computational efficiency. EfficientNet, known for its scaling capabilities in network width, depth, and resolution, achieves state-of-the-art performance with fewer parameters and reduced training time [14]. ResNet and its variant ResNeXt continue to serve as foundational architectures in medical imaging, with ResNeXt enhancing the balance between accuracy and computational complexity through grouped convolutions, making it particularly effective for AMD detection [4].

Siamese Networks for Sequential Image Analysis. Siamese networks are particularly effective for tracking disease progression in AMD by comparing consecutive OCT scans to detect changes in neovascular activity. When combined with advanced CNN architectures, Siamese networks excel in identifying subtle differences in OCT images, which are critical for timely treatment adjustments [7].

Data Augmentation Techniques. Robust data augmentation is crucial for the success of AI models in medical imaging. Techniques like random rotations, flips, and color jittering are standard, but more advanced methods such as Mixup, Cutmix, and RandAugment have been tested showing not a significant improvement [16]. Thus, only some standard data augmentation methods have been adopted, creating diverse training datasets, and enabling models to better identify subtle variations in medical images.

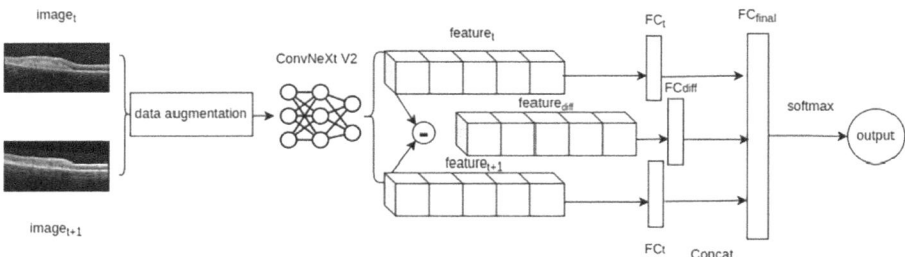

Fig. 1. The pipeline for the **OCT-DiffNet** algorithm. In this architecture, **FC** represents a fully connected layer sequence comprising `Linear + ReLU + Linear`. The data augmentation block includes essential transformations like resizing and applying other augmentations to enhance model generalization. OCT images from consecutive time points (`image`$_t$ and `image`$_{t+1}$) are fed into a shared ConvNeXt V2 backbone, extracting deep features. These features are then processed individually (`feature`$_t$ and `feature`$_{t+1}$) and compared to compute the absolute difference feature vector (`feature`$_{diff}$). Finally, these feature vectors are concatenated and passed through additional fully connected layers to produce the final classification output, indicating the disease progression.

3 Methods

The primary objective of this study is to develop a robust AI-driven framework capable of detecting the progression of neovascular activity in AMD by analyzing pairs of sequential OCT images. The focus is on identifying subtle changes between consecutive scans, which are indicative of disease worsening or improvement. By leveraging advanced deep learning architectures, the proposed method aims to provide a more accurate and timely assessment of disease progression, thereby facilitating better-informed treatment decisions for patients undergoing anti-VEGF therapy (see Fig. 1).

3.1 Data Preparation and Augmentation

In this study, we utilized a dataset of OCT images for the analysis of AMD. Each image has an uncertain resolution is provided in grayscale. The dataset is split into 80% for training and 20% for validation, ensuring class balance across the subsets. To enhance model generalization and robustness, extensive data augmentation techniques were applied to the training set. The augmentation included:

- **Color Jittering:** Adjusting the brightness and contrast of images with a probability of 25%, adding variability in illumination conditions.
- **Gaussian Blur:** Applying a Gaussian blur with a 15% probability, simulating the variability in image sharpness.
- **Random Affine Transformations:** Introducing random rotations, scaling, and translations to simulate various orientations and perspectives of the eye.
- **Horizontal Flipping:** Randomly flipping images horizontally with a 30% probability to augment symmetry in the data.
- **Normalization:** Standardizing the pixel intensity values to a mean of 0.5 and a standard deviation of 0.5, is essential for stabilizing the training process.

The validation set was subjected only to basic resizing and normalization to maintain consistency.

3.2 Model Architecture

Our proposed framework is constructed upon a modified ConvNeXt V2-Large architecture, which is pre-trained on the ImageNet dataset [3] to leverage the benefits of transfer learning. Several key modifications were introduced to tailor the architecture specifically for the analysis of AMD through OCT images.

First, the model was adapted to process single-channel (grayscale) OCT images by modifying the initial convolutional layer to accommodate a single input channel. This adaptation ensures that the model is optimized for the grayscale nature of the OCT data.

During the feature extraction phase, the original fully connected (FC) layer of the ConvNeXt V2 architecture was removed, allowing the deep features extracted from the convolutional layers to be processed by custom-designed layers tailored for detecting AMD progression.

A significant enhancement in the model is the integration of a Siamese network structure, which processes two consecutive OCT images (image_t and image_{t+1}) to capture temporal changes indicative of disease progression or regression. This Siamese network, referred to as **OCT-DiffNet**, enhances the model's ability to focus on temporal differences between the images, thereby accelerating convergence during training.

The architecture of OCT-DiffNet incorporates several critical components:

Feature Extraction with ConvNeXt V2: Both input images, image_t and image_{t+1}, are passed through the ConvNeXt V2-Large model to extract deep feature representations. The feature vectors for the images are denoted as x_t and x_{t+1}, respectively:

$$x_t = \text{ConvNeXtV2}(\text{image}_t), \quad x_{t+1} = \text{ConvNeXtV2}(\text{image}_{t+1})$$

Difference Computation: The model then computes the absolute difference between the feature vectors obtained from the two consecutive OCT images:

$$x_{\text{diff}} = |x_{t+1} - x_t|$$

Here, x_{diff} encapsulates the temporal changes between the two images, which are essential for evaluating disease progression.

Processing with Fully Connected Layers: Each feature vector (x_t and x_{t+1}) and the computed difference vector (x_{diff}) are passed through dedicated fully connected (FC) layers for further processing:

$$x'_t = \text{FC}_t(x_t), \quad x'_{t+1} = \text{FC}_t(x_{t+1}), \quad x'_{\text{diff}} = \text{FC}_{\text{diff}}(x_{\text{diff}})$$

The custom-designed layers in OCT-DiffNet are specifically tailored to capture progression-related features from individual OCT images as well as their temporal differences. This design ensures that the model focuses on both spatial characteristics and temporal dynamics crucial for detecting subtle AMD progression patterns

Combined Feature Analysis: The processed feature vectors are concatenated to form a combined feature vector:

$$x_{\text{combined}} = \text{concat}(x'_t, x'_{t+1}, x'_{\text{diff}})$$

This concatenated feature vector is then passed through a final fully connected layer to produce the output classification:

$$\text{output} = \text{FC}_{\text{final}}(x_{\text{combined}})$$

This architectural design enables **OCT-DiffNet** to effectively highlight subtle differences between consecutive OCT scans, providing a robust foundation for detecting the progression of AMD and informing treatment strategies.

3.3 Model Training

The model was trained using Cross-Entropy Loss, a standard choice for multi-class classification tasks. The Adam optimizer [6] was selected for its ability to efficiently handle sparse gradients, ensuring effective parameter updates throughout the training process. A StepLR scheduler was implemented to gradually decrease the learning rate, aiding in the stabilization of training as it progresses. Additionally, Automatic Mixed Precision (AMP) was utilized to accelerate training and reduce memory usage, enabling faster and more resource-efficient training.

3.4 Ensemble Method

To enhance the robustness and accuracy of the model, a 5-fold cross-validation approach was adopted, generating five distinct models. These models were then combined using a majority voting ensemble method, where the final prediction for each image pair was determined by the most frequently predicted class across all models. In cases where predictions were evenly split, a pre-determined reference model, trained on the entire dataset, was used to resolve the tie by providing the final classification.

3.5 Model Evaluation

The model's performance was thoroughly evaluated using several key metrics to ensure its reliability in clinical applications. These included the F1-Score, which balances precision and recall, making it particularly useful for imbalanced class distributions; Spearman's Rank Correlation, which measures the rank-based relationships between predicted and actual outcomes; Specificity, crucial for minimizing false positives in clinical settings; and a Mean Metric, which provides a composite score that summarizes overall model performance.

During training, model parameters were saved whenever there was an improvement in the mean metrics, ensuring the best-performing model was retained. The final model selection was based on its performance on the validation set. To further enhance robustness, a majority voting ensemble method was employed across five cross-validation folds. In cases of tied votes, a pre-determined reference model, trained on the entire dataset, was used to ensure consistent predictions.

4 Results

4.1 Impact of Different Models

The impact of various model architectures on performance was systematically evaluated by training ConvNeXt [10], ConvNeXt V2 [15], ResNet101 [4], ResNet50, EfficientNet [14], Vision Transformer [1], and Swin Transformer [9] models under identical training conditions (Table 1).

ConvNeXt V2 Large clearly outperformed all other models, achieving the highest scores across all metrics, particularly in F1-Score, Spearman's Rank Correlation, and Specificity. The architectural enhancements in ConvNeXt V2 Large, such as improved convolutional operations and more effective handling of high-dimensional image data, likely contribute significantly to its superior performance. This model's ability to capture complex patterns in OCT images makes it especially well-suited for the challenging task of analyzing AMD progression, underscoring its potential for clinical application.

Table 1. Results of performance metrics evaluation across different model architectures

Model	F1	Rk	Sp	Mean
ResNet50	0.7973	0.6358	0.8432	0.7581
ResNet101	0.8090	0.6511	0.8598	0.7740
Vision Transformer	0.8154	0.6623	0.8680	0.7813
Swin Transformer	0.8143	0.6611	0.8659	0.7804
ConvNeXt	0.8200	0.6642	0.8799	0.7880
EfficientNet	0.8307	0.6588	0.8812	0.7902
ConvNeXt V2 Large	**0.8454**	**0.6694**	**0.9000**	**0.8049**

4.2 Ablation Study: Impact of Siamese Network and Data Augmentation

The ablation study indicates that combining the Siamese network with data augmentation provides the highest performance across all metrics. This combination leads to significant improvements, particularly in F1-Score and Specificity. Both the Siamese network and data augmentation individually contribute to better model performance, but their synergistic effect produces the most substantial gains, demonstrating the efficacy of integrating these components in the proposed method (Table 2).

Table 2. Ablation study results evaluating the impact of the Siamese Network (SN) and Data Augmentation (DA) on model performance. Configurations tested include: No SN/DA, SN only, DA only, and both SN+DA.

Configuration	F1	Rk	Sp	Mean
No SN, No DA	0.7712	0.6511	0.7332	0.7185
SN Only	0.8246	0.7023	0.7878	0.7716
DA Only	0.8135	0.6934	0.7712	0.7594
SN + DA	**0.8454**	**0.6694**	**0.9000**	**0.8049**

5 Discussion

In this study, we explored the effectiveness of various deep learning models and data augmentation strategies for the task of detecting neovascular activity in AMD using OCT images. Our results demonstrate that both model architecture and data augmentation play critical roles in improving performance on this challenging task.

The ConvNeXt V2-Large model consistently outperformed other architectures, including ResNet50, ResNet101, EfficientNet, Vision Transformer, and

Swin transformer. This suggests that ConvNeXt V2's design, which incorporates advanced convolutional mechanisms and efficient handling of large-scale image data, is particularly well-suited for the fine-grained analysis required in AMD progression detection.

Data augmentation was another key factor in enhancing model performance. By introducing variability in the training data, particularly through techniques like ColorJitter, GaussianBlur, and RandomAffine transformations, the model was better equipped to generalize to unseen data. Our experiments confirmed that proper data augmentation yielded the best results across all metrics, including F1-Score, Spearman's Rank Correlation, and Specificity.

The ablation study further highlighted the importance of our proposed Siamese network architecture, which leverages the difference between consecutive OCT scans to more accurately track disease progression. The combination of this architecture with data augmentation resulted in the highest performance, indicating that both elements are crucial for maximizing accuracy in this context.

One of the significant contributions of this study is the introduction of a novel Siamese network, OCT-DiffNet, specifically designed for this task. The network's unique ability to focus on the differences between pairs of images accelerates convergence and improves sensitivity to subtle changes in the retinal structure. This innovation has the potential to significantly improve the clinical utility of OCT imaging in monitoring AMD, especially in patients undergoing anti-VEGF therapy.

Despite these promising results, there are limitations to our study. For instance, the reliance on synthetic data augmentation may not fully capture the real-world variability seen in clinical settings. Additionally, while our model performed well in the experimental setup, further validation on larger and more diverse datasets is necessary to confirm its generalizability.

6 Conclusion

This work presents a comprehensive evaluation of various deep learning models and data augmentation techniques for the detection of neovascular activity in AMD using OCT images. The results underscore the importance of model selection, with ConvNeXt V2 emerging as the most effective architecture. Additionally, our novel Siamese network, OCT-DiffNet, combined with aggressive data augmentation, significantly enhances the model's ability to detect subtle changes in OCT scans, thus improving the monitoring and treatment planning for patients with AMD.

Our approach achieved impressive results, including a second-place finish in the preliminary round of the MICCAI MARIO TASK1 challenge, validating its potential in a competitive setting. Future work will focus on further refining the model, particularly through the integration of additional clinical data and the exploration of more advanced ensemble methods. Ultimately, we aim to develop a robust tool that can be deployed in clinical practice to aid in the early detection and management of AMD, thereby contributing to better patient outcomes. The code is available at: https://github.com/YIDING4869/MARIO2024.

References

1. Alexey, D.: An image is worth 16x16 words: Transformers for image recognition at scale. arXiv preprint arXiv: 2010.11929 (2020)
2. Bhuiyan, A., Wong, T.Y., Ting, D., Govindaiah, A., Souied, E.H., Smith, R.T.: Artificial intelligence to stratify severity of age-related macular degeneration (AMD) and predict risk of progression to late AMD. Transl. Vis. Sci. Technol. **9**(2), 25–25 (2020)
3. Deng, J., et al.: ImageNet: a large-scale hierarchical image database. In: 2009 IEEE Conference on Computer Vision and Pattern Recognition, pp. 248–255. IEEE (2009)
4. He, K., Zhang, X., Ren, S., Sun, J.: Deep residual learning for image recognition. In: Proceedings of the IEEE Conference on Computer Vision and Pattern Recognition, pp. 770–778 (2016)
5. Jonas, J.B., Cheung, C., Panda-Jonas, S.: Updates on the epidemiology of age-related macular degeneration. Asia Pac. J. Ophthalmol. **6**(6), 493–497 (2017)
6. Kingma, D.P.: Adam: A method for stochastic optimization. arXiv preprint arXiv:1412.6980 (2014)
7. Koch, G., Zemel, R., Salakhutdinov, R., et al.: Siamese neural networks for one-shot image recognition. In: ICML Deep Learning Workshop, Lille, vol. 2, pp. 1–30 (2015)
8. Lim, L.S., Mitchell, P., Seddon, J.M., Holz, F.G., Wong, T.Y.: Age-related macular degeneration. Lancet **379**(9827), 1728–1738 (2012)
9. Liu, Z., et al.: Swin transformer: hierarchical vision transformer using shifted windows. In: Proceedings of the IEEE/CVF International Conference on Computer Vision, pp. 10012–10022 (2021)
10. Liu, Z., Mao, H., Wu, C.Y., Feichtenhofer, C., Darrell, T., Xie, S.: A convnet for the 2020s. In: Proceedings of the IEEE/CVF Conference on Computer Vision and Pattern Recognition, pp. 11976–11986 (2022)
11. Mitchell, P., Liew, G., Gopinath, B., Wong, T.Y.: Age-related macular degeneration. Lancet **392**(10153), 1147–1159 (2018)
12. Schmidt-Erfurth, U., et al.: Machine learning to analyze the prognostic value of current imaging biomarkers in neovascular age-related macular degeneration. Ophthalmol. Retina **2**(1), 24–30 (2018)
13. Spaide, R.F., Fujimoto, J.G., Waheed, N.K.: Image artifacts in optical coherence tomography angiography. Retina **35**(11), 2163–2180 (2015)
14. Tan, M.: Efficientnet: Rethinking model scaling for convolutional neural networks. arXiv preprint arXiv:1905.11946 (2019)
15. Woo, S., et al.: Convnext v2: co-designing and scaling convnets with masked autoencoders. In: Proceedings of the IEEE/CVF Conference on Computer Vision and Pattern Recognition, pp. 16133–16142 (2023)
16. Zhong, Z., Zheng, L., Kang, G., Li, S., Yang, Y.: Random erasing data augmentation. In: Proceedings of the AAAI Conference on Artificial Intelligence, vol. 34, pp. 13001–13008 (2020)

Efficient Deep Learning Models for Evaluating the Progression of Age-Related Macular Degeneration Through Optical Coherence Tomography

Abdul Qayyum[1(✉)], Moona Mazher[2], Imran Razzak[3], and Steven A. Niederer[1]

[1] Faculty of Medicine, National Heart and Lung Institute, Imperial College London, London, UK
engr.qayyum@gmail.com
[2] Centre for Medical Image Computing, Department of Computer Science, University College London, London, UK
[3] School of Computer Science and Engineering, University of New South Wales, Sydney, Australia

Abstract. Age-related macular degeneration (AMD) is a significant cause of vision impairment in older adults. While it doesn't lead to total blindness, the loss of central vision can make everyday tasks like recognizing faces, reading, driving, and detailed work much more challenging. The progression of AMD varies; for some, it advances gradually, while in others, it can worsen more quickly. In its early stages, AMD may not noticeably affect vision, which is why regular eye exams are essential for early detection and management. Identifying AMD early is key to preserving vision. Deep learning (DL) has emerged as a powerful tool for detecting early lesions and monitoring disease progression in retinal images. It can objectively identify structural changes, stage diseases, and locate specific retinal lesions. In this work, we present a deep transfer learning based MobileNetV3 framework to evaluate the AMD progression through Optical Coherence Tomography. To further improve the performance, we applied different test-time data augmentation techniques which showed significantly better performance for disease evolution and disease progression. We also explored various pretrained models, including DenseNet201, ResNet-34, and DinoV2. However, MobileNetV3 combined with test-time augmentation techniques delivered superior performance, particularly in Task 2, and performed well in Task 1. Our proposed approach achieved F1 scores of 0.727 for Task 2 and 0.716 for Task 1 on the initial leaderboard test set. On the external test dataset, the model's final overall ranking showed an F1 score of 0.769 for Task 1 and 0.217 for Task 2.

Keywords: Macular Degeneration · Disease Progression · OCT · Aging-Related Disease

1 Introduction

Age-related macular degeneration (AMD) is a leading cause of irreversible vision loss among individuals aged 50 and older. This degenerative condition primarily affects the macula, the central region of the retina, resulting in the gradual deterioration of central vision, which is essential for activities such as reading, driving, and recognizing faces. AMD is broadly classified into two stages: early AMD, characterized by the presence of drusen and pigmentary changes, and late AMD, which can manifest as either geographic atrophy (dry AMD) or neovascular AMD (wet AMD). Understanding and monitoring the progression of AMD is crucial for timely intervention and management. Optical Coherence Tomography (OCT) has emerged as a transformative imaging modality in the diagnosis and monitoring of AMD. OCT provides high-resolution, cross-sectional images of the retina, allowing clinicians to visualize the retinal layers and detect subtle changes in the macular structure that are indicative of AMD progression. Unlike traditional imaging techniques, OCT enables the detection of early anatomical changes before significant visual symptoms appear, making it a vital tool in both clinical and research settings. Numerous algorithms have been developed to automatically detect AMD using various imaging techniques. Many focus on segmenting and counting drusen and drusen-like deposits to identify the disease in its early stages. For instance, Yildirim et al. trained a U-Net deep learning (DL) model to identify early AMD biomarkers in OCT images, demonstrating high accuracy and the potential to streamline AMD screening through automated patient selection [3]. Morelle et al. presented DL-based OCT segmenter that accurately quantified drusen load by analyzing layer positions, achieving results closely aligned with expert human readers and improving on previous methods [1]. Saha et al. evaluated different pretrained DL algorithms for diagnosing AMD by detecting and classifying hyperreflective foci, hyporeflective foci within drusen, and subretinal drusenoid deposits from OCT B-scans [2]. Despite these efforts, deep learning methods may still fail to handle unpredictable changes in the data distribution. Test-Time Augmentation (TTA) is a widely used technique to enhance the accuracy of deep learning models during inference. TTA mitigates a common limitation in deep learning pipelines, where training datasets represent only a small fraction of all possible inputs. This paper aims to explore the progression of age-related macular degeneration as observed through optical coherence tomography. By analyzing OCT findings, we can gain deeper insights into the structural changes associated with different stages of AMD, leading to improved understanding, early diagnosis, and potentially more effective treatment strategies. To evaluate the progression of ageing related macular degeneration, we applied test-time data augmentation techniques and presented mobilenetv3 based architecture with transfer learning for significantly better performance for disease evolution and disease progression.

2 Proposed Method

In this work, we have used MobileNetV3 [4] based architecture aided with test-time augmentation. MobileNet V3, a lightweight model well suited for mobile devices, was designed through platform aware network architecture search and NetAdapt algorithm.

We adapted a pre-trained MobileNetV3 which is based on combination of different layers as building blocks to build the most effective models. To overcome the limitations of MobileNet V3, we adopted simple yet efficient strategy (test time augmentation) which yields transformed versions of the image for prediction and the results are merged and leverages multiple predictions to increase the model's confidence. Test-time augmentation (TTA (flips, crops, scales and intensity transforms) involves four key steps: augmentation, prediction, reversal, and merging. First, we apply the same augmentations to the test image that were used during training. Predictions are then made on both the original and the augmented images. Next, we reverse the transformations applied to the augmented predictions to bring them back to their original orientation, a process known as dis-augmentation. For instance, if a prediction was made on a flipped or rotated image, we realign the prediction to match the original image orientation. We used an 80% training and 20% validation split, employing a 5-fold cross-validation approach to train and optimize the proposed model. The overall block diagram of the proposed solution is depicted in Fig. 1.

2.1 Training Process and Fine-Tuning

We fine-tuned MobileNetV3, pretrained on the ImageNet dataset, to the specific tasks of image analysis. Here's how we approached the training process.

2.1.1 Dataset Preparation and Normalization

The images for task 1 and task 2 were preprocessed to standardize their size and resolution, ensuring compatibility with the input dimensions expected by MobileNetV3. We have resized the input images 256×256 to train the proposed model. Data augmentation techniques were applied during training to enhance the model's robustness, including z-normalization, random rotations, flips, brightness adjustments, and contrast variations [5, 6].

2.1.2 Transfer Learning

We utilized transfer learning by initializing MobileNetV3 with weights pretrained on ImageNet. This approach leveraged the general feature extraction capabilities learned from a large, diverse dataset and adapted them to the specific characteristics of OCT images. The network's deeper layers, which contain more task-specific features, were fine-tuned with a lower learning rate to avoid catastrophic forgetting of the pretrained features.

2.1.3 Optimization

We utilized the Adam optimizer along with a learning rate scheduler that reduces the learning rate when the validation loss plateaus, which helped the model converge more efficiently while minimizing overfitting. For classification tasks, we employed cross-entropy loss and applied regularization techniques like dropout and L2 weight regularization to improve generalization. The learning rate was set to 0.0001, with exponential

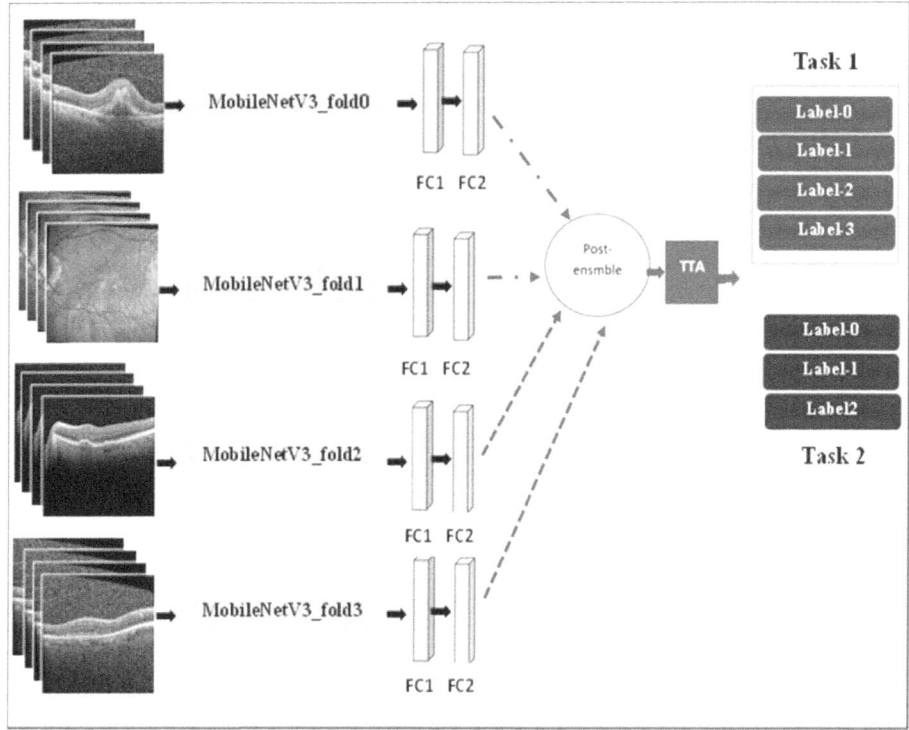

Fig. 1. The proposed approach for classification of AMD evolution and prediction of future AMD progression

decay scheduling applied during training and optimization. The model was trained for 200 epochs, and PyTorch was used for the implementation and training of the proposed model [7, 8].

2.1.4 Evaluation and Validation

The performance of MobileNetV3 was assessed using a training dataset, which was split into training and validation sets through a 5-fold cross-validation approach. We monitored metrics including accuracy, precision, recall, and F1-score to evaluate the model's performance across various classes.

2.1.5 Test-Time Augmentation (TTA)

To further enhance performance, Test-Time Augmentation (TTA) was applied. During inference, each test image was augmented multiple times (e.g., with rotations, flips, and brightness changes), and the model's predictions were averaged across these augmented versions. This technique helped mitigate the effects of minor image variations and improved the robustness of the predictions.

2.2 Ensemble Method

We trained and combined predictions from five different variations using 5-fold cross validation of MobileNetV3 models. Each variation might differ in terms of initialization, hyperparameters, or even data augmentation techniques used during training. Each model learns slightly different representations of the data, leading to a more robust and generalized final prediction. Combining predictions helps in canceling out individual model errors, improving overall accuracy and stability. Typically, the outputs of the models in the ensemble are averaged or combined using a weighted sum. This final output is used for predictions. We conducted an ablation study by comparing the performance of 'MobileNetV3_base' without any ensembling or TTA (Test-Time Augmentation) and 'MobileNetV3_ensemble,' which utilizes an ensembling approach.

2.3 MobileNetV3

MobileNetV3 is the latest iteration in the MobileNet series, specifically designed to optimize performance for mobile and edge devices while maintaining high accuracy. It builds upon the principles introduced in MobileNetV1 and MobileNetV2, incorporating advancements that enhance both efficiency and effectiveness. MobileNetV3 comes in two versions: MobileNetV3-Large and MobileNetV3-Small, each tailored for different levels of resource constraints. Here, we'll focus on the general architecture and key components that make MobileNetV3 particularly effective. The block diagram for the MobileNetV3 module is illustrated in Fig. 2.

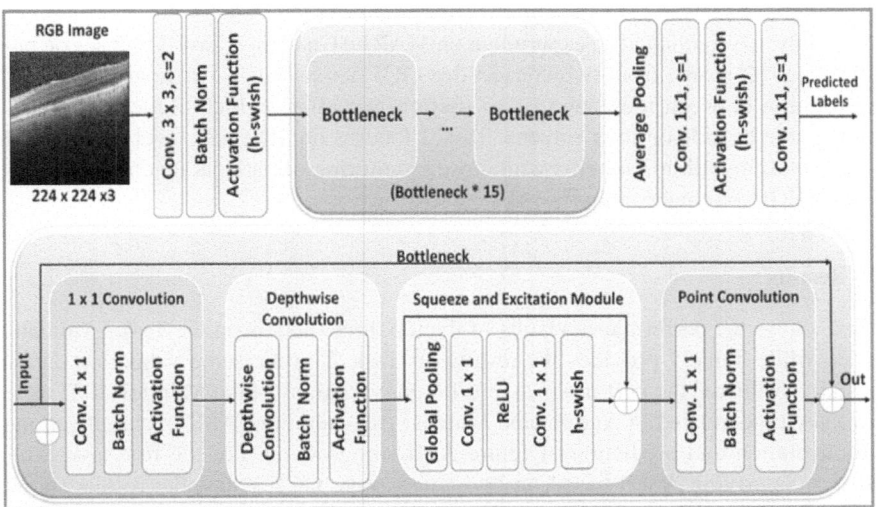

Fig. 2. Proposed MobileNetV3 block diagram used in Task1 and Task2.

MobileNetV3 consisted of inverted residuals reduce the computational burden by applying convolutions in a high-dimensional space (expanding the input channels) and then projecting back to a low-dimensional space. The linear bottleneck layer helps maintain the expressiveness of the network while reducing the dimensionality of the feature space, which is critical for efficient computation. Squeeze-and-Excitation modules are integrated into the network to adaptively recalibrate channel-wise feature responses. This is achieved by weighing each channel according to its importance, which enhances the model's ability to focus on relevant features. In MobileNetV3, SE modules are selectively applied in certain layers, balancing the trade-off between complexity and accuracy. While previous versions used ReLU as the activation function, MobileNetV3 introduces Hard-Swish (H-Swish), which is a computationally efficient approximation of the Swish activation function. This activation function retains the smoothness of Swish while being easier to compute, especially on low-power devices. MobileNetV3 employs a compound scaling strategy where the depth, width, and resolution of the network are scaled uniformly. This allows for more flexibility in optimizing the network for various deployment scenarios, depending on the available computational resources. The final stages of MobileNetV3 include a lightweight head that consists of a pooling layer followed by a fully connected layer. This design is optimized for efficiency while ensuring that the network captures high-level features effectively. A significant portion of MobileNetV3's design was informed by Neural Architecture Search (NAS), which systematically explores different network configurations to identify the most efficient structure for mobile devices.

3 Experiment

3.1 Dataset

In this study, we conducted an experiment on MARIO Challenge Task 1 focuses on pairs of 2D slices (B-scans) from two consecutive OCT acquisitions. The goal is to classify the evolution between these two slices (before and after), which clinicians typically examine side by side on their screens. Task 2 focuses on 2D slices level to predict the future evolution within 3 months with close monitoring of patients that are enrolled in an anti-VEGF treatments plan [9–14].

3.2 Results

Table 1 shows the performance results of the top three models on the Task 1 validation dataset, while Table 2 provides the results for Task 2 on the same validation dataset. The study evaluated the performance of different variations of the MobileNetV3 model in two key tasks related to Age-related Macular Degeneration (AMD): classification of AMD evolution and prediction of future AMD progression. For the first task, which involved classifying AMD based on two consecutive OCT scans, the results showed that the baseline MobileNetV3 model achieved an F1 score of 0.698, RK Generation of 0.339, and specificity of 0.837. However, when ensembling techniques were applied (MobileNetV3_ensemble), the performance improved, resulting in an F1 score of 0.716, RK Generation of 0.395, and specificity of 0.850. The best performance was observed when both ensembling and Test Time Augmentation (TTA) were used together

(MobileNetV3_ensemble + TTA), achieving an F1 score of 0.727, RK Generation of 0.391, and specificity of 0.839, indicating a more accurate overall classification.

In the second task, which involved predicting future AMD evolution, the baseline model showed moderate results with an F1 score of 0.695, but the RK Generation (−0.0035) and Quadratic-Weighted Kappa (QWK) (−0.01998) were poor, indicating limitations in predicting progression accurately. A low but non-zero score might suggest the model correctly predicted rare classes more often than expected. The ensemble model slightly improved RK Generation but still had a negative Kappa, suggesting that ensembling alone was insufficient. However, the combination of ensembling and TTA in the MobileNetV3_ensemble + TTA model led to the best performance, with an F1 score of 0.716, RK Generation of 0.063, and a positive Quadratic-Weighted Kappa of 0.0659. These results highlight that the use of advanced techniques like ensembling and TTA significantly enhances the model's performance, particularly in complex tasks like predicting the progression of AMD.

Table 1. Classification of AMD evolution (Task1)

Model	F1 score	RK Generation	Specificity
MobileNetV3_base	0.698	0.339	0.837
MobileNetV3_ensemble	0.716	0.395	0.850
MobileNetV3_ensemble + TTA	0.727	0.391	0.839

Table 2. Future evolution prediction (Task2).

Model	F1 score	RK Generation	Specificity	QWK
MobileNetV3_base	0.695	−0.0035	0.667	−0.01998
MobileNetV3_ensemble	0.668	0.01374	0.665	−0.0335
MobileNetV3_ensemble + TTA	0.716	0.063	0.6765	0.0659

In this study, we focused on improving the performance of the MobileNetV3 model for two critical tasks related to Age-related Macular Degeneration (AMD): classification of AMD evolution and prediction of future AMD progression. We utilized the MobileNetV3 model with a 100% pretrained backbone, fine-tuning it specifically for our task 1 and task 2 datasets. This approach allowed us to leverage the powerful feature extraction capabilities of the pretrained model while adapting it to the specific nuances of the AMD datasets.

For task 2, which aimed at predicting future AMD evolution, the MobileNetV3_ensemble + TTA model again outperformed the other variations, achieving not only higher F1 scores and specificity but also a positive Quadratic-Weighted Kappa. This is particularly significant as the Kappa metric reflects the model's ability to correctly predict progression stages, which is crucial for clinical decision-making. The positive Kappa value underscores the model's enhanced agreement with actual disease progression, suggesting that the ensemble and TTA approach mitigated overfitting and improved prediction stability.

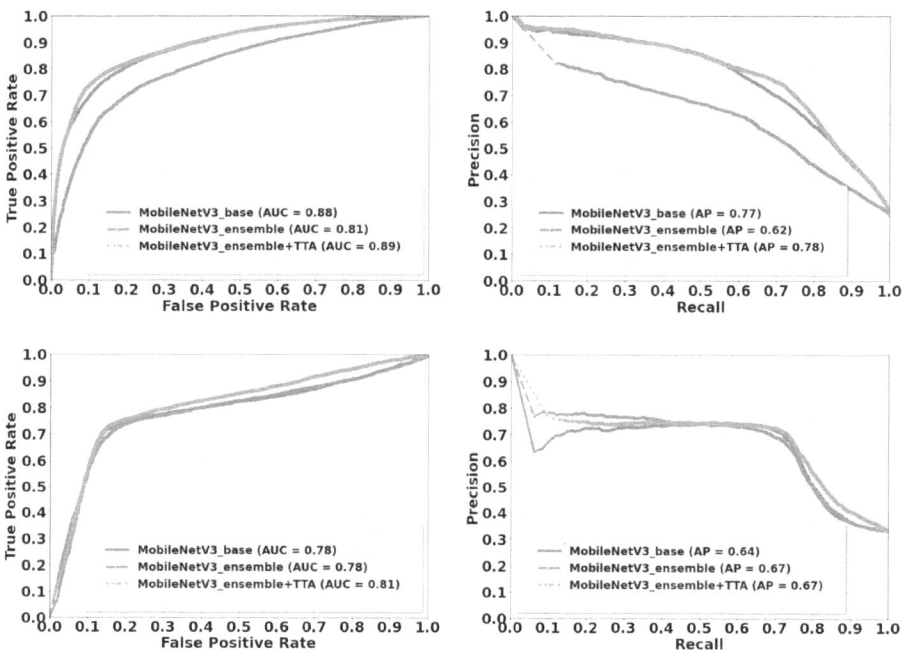

Fig. 3. The ROC and precision recall curves using proposed model for task1 and task2.

In Fig. 3, the first row presents both the ROC curves and Precision-Recall (PR) curves for Task 1, while the second row illustrates these curves for Task 2 using validation dataset. Across both tasks, the proposed MobileNetV3 model, incorporating ensembling and test-time augmentation (TTA), consistently outperforms the other model variations. Specifically, MobileNetV3 with ensembling and TTA demonstrates superior performance with a more optimal ROC curve and PR curve compared to the MobileNetV3 ensemble without TTA and the base MobileNetV3 model without any ensembling. In Task 1, this improvement is evident as the ROC curve for the augmented ensemble model is closer to the top-left corner of the plot, indicating a higher true positive rate at lower false positive rates. Similarly, the PR curve shows higher precision across different recall levels, reflecting the model's ability to maintain accuracy even as the recall increases. For Task 2, a similar trend is observed. The MobileNetV3 model with ensembling and TTA again exhibits superior ROC and PR curves, suggesting that combining these techniques provides the model with more robustness and generalization capability. This improvement is critical in complex tasks where balancing precision and recall is essential for reliable performance. Overall, the consistent patterns seen across both tasks underline the effectiveness of leveraging ensembling and test-time augmentation in enhancing model performance, particularly in scenarios requiring robust classification under varying conditions.

4 Conclusion

In this work, we introduced a MobileNetV3-based framework specifically designed to evaluate the progression of Age-related Macular Degeneration (AMD). Our approach stands out due to its ability to outperform other architectures, particularly when advanced data augmentation techniques, such as Test Time Augmentation (TTA), are integrated. Through comprehensive experiments on two key tasks, classification of AMD evolution (task 1) and prediction of future AMD progression (task 2. Our framework demonstrated superior performance. The MobileNetV3-based model demonstrated strong performance, achieving an impressive F1 score of 0.727 for Task 1 and 0.716 for Task 2 on the initial leaderboard test dataset, underscoring its robustness and accuracy in managing complex medical imaging data.

These results underscore the effectiveness of combining a lightweight yet powerful architecture like MobileNetV3 with data augmentation strategies. The significant improvement in performance metrics reflects the model's enhanced ability to generalize and accurately capture the intricate patterns associated with AMD progression. This success not only validates the potential of MobileNetV3 for such tasks but also points to the critical role of augmentation techniques in boosting the performance of deep learning models in medical imaging. Our framework, therefore, offers a promising tool for clinicians in making more accurate and timely assessments of AMD progression, ultimately contributing to better patient outcomes.

References

1. Morelle, O., Wintergerst, M.W., Finger, R.P., Schultz, T.: Accurate drusen segmentation in optical coherence tomography via order-constrained regression of retinal layer heights. Sci. Rep. **13**(1), 8162 (2023)
2. Saha, S., et al.: Automated detection and classification of early amd biomarkers using deep learning. Sci. Rep. **9**(1), 10990 (2019)
3. Yildirim, K., et al.: U-net-based segmentation of current imaging biomarkers in oct-scans of patients with age related macular degeneration. In: Caring is Sharing–Exploiting the Value in Data for Health and Innovation, pp. 947–951. IOS Press (2023)
4. Howard, A., et al.: Searching for mobilenetv3. In: Proceedings of the IEEE/CVF International Conference on Computer Vision, pp. 1314–1324 (2019)
5. Qayyum, A., Benzinou, A., Mazher, M., Meriaudeau, F.: Efficient multi-model vision transformer based on feature fusion for classification of dfuc2021 challenge. In Diabetic Foot Ulcers Grand Challenge, pp. 62–75. Springer International Publishing, Cham (2021)
6. Li, X., et al.: The state-of-the-art 3D anisotropic intracranial hemorrhage segmentation on non-contrast head CT: The INSTANCE challenge. arXiv preprint arXiv:2301.03281 (2023)
7. Qayyum, A., Razzak, I., Tanveer, M., Mazher, M., Alhaqbani, B.: High-density electroencephalography and speech signal based deep framework for clinical depression diagnosis. In: IEEE/ACM Transactions on Computational Biology and Bioinformatics 20, no. 4, pp. 2587–2597 (2023)

8. Nan, Y., et al.: Hunting imaging biomarkers in pulmonary fibrosis: benchmarks of the AIIB23 challenge. Med. Image Anal. **97**, 103253 (2024)
9. Jonas, J.B., Cheung, C.M.G., Panda-Jonas, S.: Updates on the epidemiology of age-related macular degeneration. Asia Pac. J. Ophthalmol. (Phila) **6**(6), 493–497 (2017)
10. Rosenfeld, P.J., Brown, D.M., Heier, J.S., Boyer, D.S., Kaiser, P.K., Chung, C.Y., et al.: Ranibizumab for neovascular age-related macular degeneration. N. Engl. J. Med. **355**(14), 1419–31 (2006)
11. Rasmussen, A., Sander, B.: Long-term longitudinal study of patients treated with ranibizumab for neovascular age-related macular degeneration. Curr. Opin. Ophthalmol. **25**(3), 158–63 (2014)
12. Freund, K.B., Korobelnik, J.-F., Devenyi, R., Framme, C., Galic, J., Herbert, E., et al.: Treat-and-extend regimens with anti- VEGF agents in retinal diseases : a literature review and consensus recommendations. Retina (Philadelphia, Pa) **35**(8), 1489–506 (2015)
13. Bhuiyan, A., Wong, T.Y., Ting, D.S.W., Govindaiah, A., Souied, E.H., Smith, R.T.: Artificial intelligence to stratify severity of age-related macular degeneration (AMD) and predict risk of progression to late AMD. Transl. Vis. Sci. Technol. **9**(2), 25 (2020). https://doi.org/10.1167/tvst.9.2.25, PMID: 32818086; PMCID: PMC7396183
14. Li, E., Donati, S., Lindsley, K.B., Krzystolik, M.G., Virgili, G.: Treatment regimens for administration of anti-vascular endothelial growth factor agents for neovascular age-related macular degeneration. Cochrane Database Syst. Rev. **5**(5), CD012208 (2020). https://doi.org/10.1002/14651858.CD012208.pub2, PMID: 32374423; PMCID: PMC7202375

Optimizing Anti-VEGF Treatment Strategies with AI-Based Neovascular AMD Detection

Yi Ding

The University of Edinburgh, Edinburgh EH8 9YL, UK
scyydd4@outlook.com

Abstract. Age-related Macular Degeneration (AMD) is a leading cause of visual impairment globally, with neovascular AMD (nAMD) posing significant challenges for effective treatment. This study presents a novel approach leveraging the Clustering-constrained Attention Multiple instance learning Single-branch (CLAM-SB) framework combined with ConvNeXt V2 for feature extraction to predict neovascular AMD (nAMD) progression from optical coherence tomography (OCT) images. Our method addresses the limitations of traditional instance-based approaches by aggregating information across multiple images, resulting in enhanced prediction stability and accuracy. Additionally, we mitigate class imbalance using a weighted random sampler, further improving model performance. The proposed approach not only demonstrates superior efficacy in monitoring disease activity but also secured first place in the preliminary round of the Monitoring Age-related Macular Degeneration Progression in Optical Coherence Tomography Task 2 at MICCAI 2024, underscoring its potential as a robust tool for individualized treatment planning in patients undergoing anti-VEGF therapy.

Keywords: Deep learning · Optical coherence tomography · Age-related macular degeneration

1 Introduction

Age-related Macular Degeneration (AMD) is a leading cause of severe visual impairment and blindness among the elderly, affecting nearly 196 million people worldwide [6]. AMD is a progressive retinal disease that primarily affects the macula, leading to the deterioration of central vision. It typically manifests in individuals aged 50 and above, with a significant increase in prevalence among those over 65 years of age [10]. Advanced stages of AMD, including geographic atrophy (GA) and neovascular AMD (nAMD), account for approximately 20% of cases and are the primary contributors to vision loss in developed countries [9].

The advent of anti-vascular endothelial growth factor (anti-VEGF) therapies has revolutionized the management of nAMD, offering the potential to slow disease progression and even improve visual outcomes [10]. However, the effectiveness of these treatments relies heavily on timely intervention and continuous monitoring to detect disease recurrence or progression. Optical coherence tomography (OCT), a non-invasive imaging technique, has become the gold standard for diagnosing and monitoring nAMD due to its ability to visualize retinal structures in high detail [3]. Accurate interpretation of OCT images is crucial for identifying exudative signs, such as subretinal and intraretinal fluid, which are indicative of neovascular activity and necessitate further treatment [3].

In recent years, artificial intelligence (AI) has shown significant promise in automating the analysis of medical images, including OCT scans, to assist in the early detection and monitoring of AMD [11]. Early efforts primarily focused on classifying and segmenting retinal images to diagnose various stages of AMD, including early/intermediate AMD (iAMD) and advanced stages like GA and nAMD [1]. Models such as ResNet have been widely used for these tasks due to their robustness and integration ease [4]. However, newer architectures like EfficientNet [12] and ConvNeXt V2 [13] have demonstrated superior performance in medical imaging tasks by optimizing both accuracy and computational efficiency.

Despite these advancements, most studies have focused on the initial diagnosis or risk prediction of AMD rather than on monitoring disease progression in patients already undergoing treatment. This gap is particularly crucial for patients receiving anti-VEGF therapy, where timely detection of disease recurrence or progression can significantly impact treatment outcomes [3]. Multi-instance learning (MIL) has gained attention as a method suitable for scenarios where labels are associated with groups of images (i.e., bags) rather than individual images [2]. MIL allows the model to aggregate information from multiple instances within a bag, making it particularly well-suited for medical imaging tasks where multiple scans or images from the same patient must be analyzed collectively [5].

To further enhance MIL models, attention mechanisms have been integrated to improve model interpretability and performance. The CLAM-SB model [8], a sophisticated MIL variant, has been designed to capture subtle changes in complex medical data by leveraging both instance-level and bag-level information. CLAM-SB has shown to outperform traditional attention-based MIL models in various applications by providing more nuanced and accurate predictions, which are crucial in a clinical setting where treatment decisions are based on longitudinal monitoring.

In this study, we aim to address the challenges of monitoring disease progression in patients with nAMD undergoing anti-VEGF therapy. We developed and evaluated machine learning models to predict the progression of nAMD, comparing traditional instance-based learning methods with advanced MIL frameworks, including the CLAM-SB model. Additionally, we explored the impact of different feature extraction architectures, such as ResNet [4], EfficientNet [12], and ConvNeXt V2 [13], on model performance. Our approach was rigorously tested in the Monitoring Age-related Macular Degeneration Progression in Optical Coherence Tomography for Task 2, part of the MICCAI 2024 challenge, where it achieved first place in the preliminary round, demonstrating superior efficacy and potential for clinical applications.

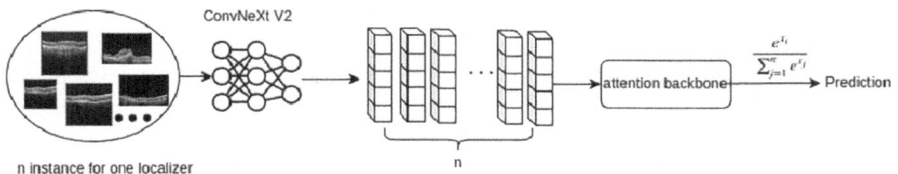

Fig. 1. Overview of the proposed multi-instance learning framework for predicting AMD progression.

2 Methodology

2.1 Objective

The objective of this study is to develop a predictive model capable of determining the likelihood of disease progression based on medical imaging data. The task involves classifying whether the condition depicted in the images is likely to deteriorate over time. To solve this problem, we proposed a new framework (see Fig. 1).

2.2 Data Preprocessing

The dataset comprises medical images, each associated with a label indicating potential disease progression. To preserve contextual information, the images were grouped by their corresponding localizers. A localizer refers to an identifier grouping multiple OCT slices from the same patient. This ensures that contextual information, such as longitudinal progression, is preserved during training. Each image was converted to grayscale and resized to a fixed dimension. The images were then normalized using dataset-specific mean and standard deviation values. To address data imbalance in the training set, we employed a weighted random sampler. This approach ensures that each class is represented proportionally during training, allowing the model to focus more on minority classes. This method contributes to more stable training and improved model performance, particularly in scenarios where some classes are underrepresented.

No data augmentation techniques were applied, as preliminary experiments indicated that augmentation could introduce instability in model training and potentially alter the key characteristics of the medical images, leading to less reliable predictions.

2.3 Feature Extraction

For feature extraction, we employed the ConvNeXt V2 architecture [13], a deep convolutional neural network pre-trained on ImageNet. To adapt the model to our specific use case, we modified the first convolutional layer to accept a single input channel, enabling the processing of grayscale images. Additionally, the final fully connected layer of the ConvNeXt V2 model was removed to generate a feature vector of size 1000 for each image, instead of producing classification outputs. These extracted features were saved for each image, optimizing the training process by avoiding redundant computations during model training.

The CLAM-SB model [8] was employed in our study due to its suitability for MIL scenarios, where each localizer or "bag" contains multiple images, referred to as instances. The model is designed to utilize a gated attention mechanism, which selectively focuses on the most relevant instances within each bag, thereby extracting and emphasizing critical features from the images. In our configuration, the CLAM-SB model was set up with an embedding dimension of 1000 and was tailored to classify images into three distinct classes, improve, stable, or worse. Additionally, the attention mechanism was implemented without dropout, which helped stabilize the model's performance during training.

2.4 Training

During the training process, the feature extractor (ConvNeXt V2) was frozen, meaning its parameters were not updated. This decision was made to leverage the pre-trained features from ImageNet, which are well-suited for general image representation tasks. By freezing the feature extractor, we significantly reduced the number of parameters to optimize, focusing solely on the parameters of the CLAM-SB model. The model's parameters were updated using the Adam [7] optimizer with a learning rate scheduler, which progressively reduced the learning rate. The model was trained for a predetermined number of epochs, during which the loss function, CrossEntropyLoss, was weighted to account for class imbalances in the dataset. The model's performance was monitored using a variety of metrics, including accuracy, F1 score, and quadratic weighted kappa (QWK).

The training process also involved a custom sampling strategy. Specifically, a WeightedRandomSampler was employed to ensure that each batch received a balanced representation of classes. This was crucial in addressing the class imbalance in the dataset.

3 Results

In this study, we conducted a series of experiments to evaluate different methods and configurations for predicting disease progression in AMD based on OCT imaging data.

3.1 Instance-Based Learning vs. Multi-instance Learning

We compared two primary approaches:

- **Instance-Based Learning with Thresholding:** In this method, each image (instance) was independently evaluated to determine the likelihood of disease progression. A threshold mechanism was applied, where disease worsening was indicated if a certain proportion of instances exceeded the threshold. However, this approach proved to be unstable and inconsistent due to its sensitivity to threshold settings.
- **MIL:** To improve stability, we adopted a MIL approach, aggregating information across multiple instances before making a final prediction. This method showed significant improvements in both the stability and accuracy of predictions, reducing the impact of noise from individual instances.

3.2 Ablation Study

Table 1. Ablation study results evaluating the impact of Data Augmentation (DA), Dropout Regularization (DO), and Top-K Selection (TK) on model performance. Configurations tested include: baseline (no DA/DO/TK), DA only, DO only, TK only, and combinations of these techniques.

Config	F1	Rk	Sp	Mean
DA Only	0.6512	0.3123	0.6453	0.5363
DO Only	0.6628	0.3241	0.6525	0.5465
TK Only (K=8)	0.6850	0.3457	0.6781	0.5696
DA + DO	0.5645	0.2389	0.6124	0.4659
DA + TK	0.6126	0.2468	0.6396	0.5000
DO + TK	0.6312	0.3694	0.7053	0.5398
No DA, No DO, No TK	**0.7954**	**0.4521**	**0.7625**	**0.6105**

Table 1 presents the results of the ablation study, evaluating the impact of data augmentation (DA), dropout regularization (DO), and top-k selection (TK) on model performance. To facilitate interpretation, the metrics used in the table are defined as follows. Rkrefers to Spearman's rank correlation coefficient, which measures the monotonic relationship between the predicted progression rankings and the ground truth rankings. A higher RL indicates better alignment between the model predictions and the actual disease progression order. Sp, on the other hand, represents specificity, which is defined as the proportion of true negative predictions among all actual negative cases. This metric reflects the model's ability to correctly identify instances with no disease progression, a critical aspect for ensuring reliable predictions in clinical applications.

Data Augmentation. We explored the effects of data augmentation techniques (like Mixup, standard data augmentation) on model performance. While augmentation is typically used to enhance model robustness, in our case, it led to instability by altering critical features necessary for accurate predictions. Consequently, data augmentation was not included in the final model.

Dropout Regularization. The impact of dropout regularization was assessed by comparing models trained with and without dropout. We found that models with dropout exhibited poorer performance, likely due to the reduction in model capacity and disruption of the attention mechanisms. Therefore, dropout was excluded from the final model configuration.

Top-K vs. Full Training. We conducted an experiment comparing models trained using the top-k most relevant instances (with $k = 16$, $k = 8$, $k = 4$, $k = 2$) versus all instances. Contrary to expectations, the model trained on the top-k instances performed worse than the model trained on the full set, indicating that leveraging the entire dataset leads to better generalization.

3.3 Weighted Loss vs. Weighted Random Sampler

We also explored methods to address class imbalance in the dataset by comparing the effects of using a weighted loss function versus a weighted random sampler. The results, shown in Table 2, indicate that while both methods improved the model's ability to handle underrepresented classes, the weighted random sampler provided slightly more stable results, likely due to its ability to ensure balanced class representation during training.

Table 2. Results between Weighted Loss Function and Weighted Random Sampler for Handling Class Imbalance on Different Performance Metrics.

Method	F1	Rk	Sp	QWK	Mean
Weighted Loss	0.7700	0.2862	0.7522	0.2832	0.5229
Weighted Sampler	**0.7954**	**0.4521**	**0.7625**	**0.4321**	**0.6105**

3.4 Comparison of Feature Extractors

The choice of feature extractor plays a critical role in the performance of deep learning models. In our experiments, we compared three popular feature extractors: ResNet-50 [4], ConvNeXt V2 [13], and EfficientNet [12]. As shown in Table 3, ConvNeXt V2 outperformed the others across all metrics, achieving the highest F1 score, Spearman correlation, and mean metrics. The superior performance of ConvNeXt V2 can be attributed to its ability to capture complex features more effectively, making it well-suited for the intricate task of detecting AMD progression.

Table 3. Results between Different Feature Extractors' Impact on Different Performance Metrics.

Method	F1	Rk	Sp	QWK	Mean
ResNet-50	0.7113	0.3757	0.6835	0.2888	0.5148
EfficientNet	0.7612	0.4163	0.7125	0.3457	0.5589
ConvNeXt-V2	**0.7954**	**0.4521**	**0.7625**	**0.4321**	**0.6105**

Table 4. CLAM-SB vs. Attention-Based MIL on Different Performance Metrics.

Method	F1	Rk	Sp	QWK	Mean
Attention-Based MIL	0.7449	0.3364	0.7157	0.3236	0.5246
CLAM-SB	**0.7954**	**0.4521**	**0.7625**	**0.4321**	**0.6105**

3.5 CLAM-SB vs. Attention-Based MIL

We also compared the performance of the CLAM-SB [8] model with a basic attention-based MIL model. Our findings demonstrated that CLAM-SB outperformed the basic attention-based model in terms of accuracy, robustness, and generalization, making it the preferred choice for our task (See Table 4).

These experiments underscore the critical role of selecting appropriate learning strategies, model architectures, and training methodologies in the context of complex medical imaging tasks like AMD progression prediction. The insights gained from these comparative analyses informed the design of our final approach, ensuring improved stability and accuracy in the prediction of disease progression.

4 Discussion

This study explored deep learning strategies for predicting the progression of Age-related Macular Degeneration (AMD) using optical coherence tomography (OCT) images, with a focus on optimizing accuracy and stability in clinical applications.

Instance-Based Learning vs. Multi-Instance Learning: We initially employed an instance-based learning approach, where each image was treated independently. However, this method exhibited significant instability, with small variations in input or decision thresholds leading to substantial fluctuations in predictions. This outcome underscored the inherent complexity of AMD progression, where subtle patterns across multiple images are crucial for accurate assessment. The transition to a MIL framework, which considers multiple images collectively, significantly improved prediction stability and performance. The MIL approach's ability to aggregate information across instances proved essential in capturing the nuanced changes associated with disease progression, as reflected in higher F1 scores and more consistent predictions.

Data Augmentation and Regularization: While data augmentation is often used to improve model generalization, our experiments revealed that it introduced instability by altering critical image features. Consequently, minimal augmentation was applied to maintain data integrity. Additionally, dropout regularization, typically used to prevent overfitting, was found to reduce model performance in this context. Removing dropout allowed the model to retain more detailed features, which are essential for accurate analysis of OCT images.

Feature Extractors and Model Architecture: Among the feature extractors tested, ConvNeXt V2 outperformed ResNet-50 and EfficientNet, achieving the highest metrics across all evaluations. This superior performance is attributed to ConvNeXt V2's ability to capture complex retinal features more effectively. Furthermore, the CLAM-SB model demonstrated significant advantages over a basic attention-based MIL model, particularly in handling the nuanced data required for accurate AMD progression prediction.

Clinical Implications: The findings of this study have direct clinical relevance, as improved prediction models can enhance patient outcomes by supporting personalized treatment plans and timely interventions. The robust performance of the MIL framework, particularly when combined with the CLAM-SB model and ConvNeXt V2 feature extractor, highlights its potential for broader application in clinical settings.

Future Directions: While this study focused on OCT images, future research should consider integrating additional clinical data, such as genetic and lifestyle factors, to develop more comprehensive predictive models. Expanding this approach to other retinal diseases or different medical conditions could further demonstrate its versatility and impact.

5 Conclusion

In this study, we developed and evaluated a predictive model for determining the trend of disease progression in patients with AMD based on OCT images. Our approach leveraged the CLAM-SB [8] MIL framework, which was shown to outperform traditional instance-based methods in both accuracy and stability. By employing advanced feature extraction techniques, such as ConvNeXt V2 [13], and addressing class imbalance through the use of a weighted random sampler, we further enhanced model performance.

Our findings indicate that multi-instance learning, coupled with careful handling of class imbalances and the use of state-of-the-art feature extraction models, provides a robust framework for predicting disease progression in AMD patients undergoing anti-VEGF treatment. The superior performance of the CLAM-SB model underscores its potential for application in clinical settings, where accurate and reliable predictions are essential for individualized treatment planning.

Future work will explore the integration of additional clinical variables and longitudinal data to further refine predictive accuracy and applicability in broader patient populations.

Our code is available at: https://github.com/YIDING4869/MARIO2.

References

1. Bhuiyan, A., Wong, T.Y., Ting, D., Govindaiah, A., Souied, E.H., Smith, R.T.: Artificial intelligence to stratify severity of age-related macular degeneration (amd) and predict risk of progression to late amd. Transl. Vis. Sci. Technol. **9**(2), 25 (2020)
2. Carbonneau, M.A., Cheplygina, V., Granger, E., Gagnon, G.: Multiple instance learning: a survey of problem characteristics and applications. Pattern Recogn. **77**, 329–353 (2018)
3. Freund, K.B., et al.: Treat-and-extend regimens with anti-vegf agents in retinal diseases: a literature review and consensus recommendations. Retina **35**(8), 1489–1506 (2015)
4. He, K., Zhang, X., Ren, S., Sun, J.: Deep residual learning for image recognition. In: Proceedings of the IEEE Conference on Computer Vision and Pattern Recognition, pp. 770–778 (2016)
5. Ilse, M., Tomczak, J., Welling, M.: Attention-based deep multiple instance learning. In: International Conference on Machine Learning, pp. 2127–2136. PMLR (2018)
6. Jonas, J.B., Cheung, C., Panda-Jonas, S.: Updates on the epidemiology of age-related macular degeneration. Asia-Pac. J. Ophthalmol. **6**(6), 493–497 (2017)
7. Kingma, D.P.: Adam: a method for stochastic optimization. arXiv preprint arXiv:1412.6980 (2014)
8. Lu, M.Y., Williamson, D.F., Chen, T.Y., Chen, R.J., Barbieri, M., Mahmood, F.: Data-efficient and weakly supervised computational pathology on whole-slide images. Nat. Biomed. Eng. **5**(6), 555–570 (2021)
9. Mitchell, P., Liew, G., Gopinath, B., Wong, T.Y.: Age-related macular degeneration. Lancet **392**(10153), 1147–1159 (2018)
10. Rosenfeld, P.J., et al.: Ranibizumab for neovascular age-related macular degeneration. N. Engl. J. Med. **355**(14), 1419–1431 (2006)
11. Sampson, D.M., Sampson, D.D.: Ai-driven innovations in signal/image processing and data analysis for optical coherence tomography in clinical applications. In: Biophotonics and Biosensing, pp. 417–480. Elsevier (2024)
12. Tan, M.: Efficientnet: rethinking model scaling for convolutional neural networks. arXiv preprint arXiv:1905.11946 (2019)
13. Woo, S., et al.: Convnext v2: co-designing and scaling convnets with masked autoencoders. In: Proceedings of the IEEE/CVF Conference on Computer Vision and Pattern Recognition, pp. 16133–16142 (2023)

Monitoring Age-Related Macular Degeneration Progression in Optical Coherence Tomography (MARIO), Task 1 - MICCAI Challenge 2024, *jkulinzstudents* Submission

Marcel Huber[✉], Patrick Binder, and Markus Frohmann

Johannes Kepler University Linz, Linz, Austria
marceljulianhuber@gmail.com

Abstract. Age-related Macular Degeneration (AMD) is a leading cause of vision loss in developed countries. The evolution of neovascular activity in AMD, particularly in response to anti-VEGF treatments, is critical for effective management. This paper presents a simple but highly effective method for classifying the evolution between two consecutive Optical Coherence Tomography (OCT) B-scans, focusing on detecting changes indicative of neovascular activity. Our approach leverages state-of-the-art deep learning techniques, namely transfer learning of the DinoV2 model with extensive data augmentations, to improve the planning of individualized anti-VEGF treatment strategies. It achieves an F1 score of 0.83 on the public leaderboard of task 1 of the MARIO challenge.

Keywords: Age-related Macular Degeneration (AMD) · Optical Coherence Tomography (OCT) · Deep Learning

1 Introduction

Age-related Macular Degeneration (AMD) is a leading cause of incurable progressive vision loss and blindness, affecting almost 200 million people worldwide. With age being the leading risk factor, the number of diseases is expected to rise. AMD involves pathologic changes in the retinal layers of the macula and progresses to increasing visual impairment. Since the disease cannot be cured, preventative and decelerating measures in the first stages require early detection. Indicators such as small drusen (retinal deposits) are often the first sign of the initial, often asymptomatic, stage of AMD [12]. For their diagnosis and evaluation, macular layers of patients are visualized with optical coherence tomography (OCT), a crucial technology for AMD treatment [11].

In recent years, machine learning (ML) has emerged as a powerful tool for analyzing medical images, including OCT scans [6]. The ability of ML algorithms to learn complex patterns has enabled new pathways for the early detection, diagnosis, and prognosis of AMD [13,14].

Furthermore, several recent advancements in anti-VEGF (vascular endothelial growth factor) therapies have significantly improved the management of disease progression, making its treatment more feasible [3,9]. Despite these advancements, the treatment plans still require timely detection of neovascular activity evolution. Hence, early identification of such changes is crucial for effective intervention in AMD. However, to date, the detection of these changes has been challenging.

To fill this gap, we propose a reliable and automated method for classifying the evolution of neovascular activity by analyzing pairs of consecutive OCT B-scans. We leverage Vision Transformers (ViTs) trained on a large, diverse set of images and add a lightweight module to classify disease progression. Our method will facilitate better treatment planning for AMD patients under anti-VEGF therapy.

2 Method

In this work, we propose the usage of ViT models [2] to identify changes between the input of two B-scans to classify its change activity. Our pre-processing pipeline, which aims to enhance relevant features, includes resizing and normalization. We use the pretrained DinoV2 with registers [1,7], a ViT model trained on a large dataset of 104M images and different objectives, and add a Multi-Layer Perceptron (MLP), taking as input the output of the DinoV2 ViT. We utilize this pipeline to capture even very subtle changes between consecutive B-scans and classify neovascular activity. We employ k-fold cross-validation with $k = 10$, training ten models using different random seeds for data stratification to ensure robust evaluation. The stratification pipeline divides patient IDs into training and validation sets. By applying varying random seeds, we generate distinct patient stratification for each fold, ensuring that every fold comprises a unique subset of patients. This approach enhances the robustness of our evaluation by leveraging the diversity inherent in multiple independent folds. Finally, we aggregate predictions from these models and post-process them as described in Sect. 2.5. Specifically, the softmax scores of the k-fold trained models are summed, and the class with the maximum cumulative score is selected as the final predicted class.

2.1 Dataset

We use a dataset of 14,497 OCT images from several consecutive medical examinations of 68 patients, as provided by the challenge organizers. Figure 1 illustrates an example of two successive OCT images from the same patient taken from the training set.

We divide our dataset into k different random folds for cross-validation and stratify based on patient ID. Each stratified split aims to maintain a train-validation ratio of 0.85:0.15. Our stratified splits ensure that the distribution of classes remains consistent across both training and validation subsets. After

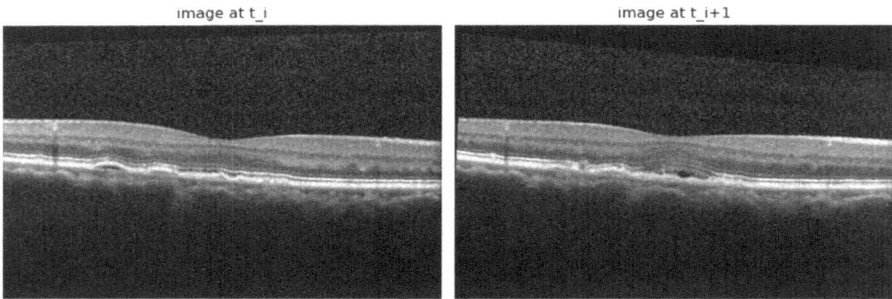

Fig. 1. OCT images of the provided dataset from the same patient over two consecutive timesteps t_i and t_{i+1}.

training and validation, our models' predictions were evaluated on the public leaderboard, which we utilized as a test set. This helped us assess the model's performance and generalizability on unseen data.

The dataset contains four target classes: *reduced* (0), *stable* (1), *worsened* (2), and *other* (3). Moreover, each scan has additional data associated with it that are related to the patient, such as their age or the time between the current and the last visit. For more details about the dataset, we refer to the challenge website.[1]

2.2 Pre-processing

For each OCT scan, we apply a series of pre-processing steps designed to make the model rely more on features relevant to neovascular activity. Inspired by data augmentations within general computer vision models and more specialized medical ones, we derive several pre-processing steps and data augmentations.

First, we resized all images to a uniform resolution of 224 × 224 pixels to (1) ensure consistency across the dataset and (2) match the input requirements of the pretrained models described in Sect. 2.4. For leaderboard submissions, we applied various data augmentation techniques, including random horizontal flipping, zooming, cropping (scale=(0.8, 1.0), ratio=(0.75, 1.333)), rotation (0-15 degrees), color jittering (brightness=0.1, contrast=0.1, saturation=0.1, hue=0.1), and retinal flattening. These augmentations aimed to enhance the model's robustness to typical variations in scan conditions.

Additionally, Gaussian noise (mean=0.0, std=0.1, p=0.5) and speckle noise (prob=0.5, intensity=0.1) were selectively introduced to simulate real-world noise conditions and improve the model's ability to distinguish pathological features, such as neovascular activity, from artifacts. While these augmentations showed promise during the leaderboard phase, limited computational resources constrained our ability to test them comprehensively.

---

[1] https://www.codabench.org/competitions/2852/.

Table 1. Performance metrics of our best submission on our training and validation sets as well as on the public leaderboard.

Metric	Train	Validation	Leaderboard
F1 Score	0.8991	0.773	0.82639
Loss	0.2904	0.6781	-
Matthew's Corr.	0.8091	0.5658	0.63445
Specificity	0.9512	0.8903	0.90113

For the final phase submission, we adopted a simpler preprocessing pipeline: resizing followed by intensity normalization (`mean=0.35, std=0.5`). The normalization values were empirically determined through experiments to maximize the contrast of regions of interest.

2.3 Proposed Method

Our method utilizes a ViT [2] to capture subtle changes between the two input B-scans. Specifically, we employ the DinoV2 model with registers (ViT-B/14 distilled[2]) with 86 million parameters [8]. Using a distilled model saves considerable computation, making our method more accessible than other methods relying on much larger models. We use this model to extract features from each preprocessed image, resulting in a feature vector of length 768 for each image. We then concatenate the feature vectors from the consecutive images *image_at_ti* and *image_at_ti+1*, producing a combined vector of length $1,536$. This vector is then fed into a two-layered MLP, which reduces the vector to 4 features corresponding to the number of classification labels within the dataset. We train $k = 10$ models, wherein for each fold, we vary the seed for stratification (c.f., 2.1). The final output of the model indicates whether there is an evolution in neovascular activity between the two input B-scans.

2.4 Pre-training

We rely on transfer learning (`lr=1e-5`) by initializing using ViT weights trained on a large dataset consisting of 141M images, thereby starting with a strong foundation in feature extraction. For details, we refer to [1] and [7].

2.5 Post-processing

After model training, we sum the softmax scores from the k-fold trained models, and the class with the maximum cumulative score is taken as the final prediction. With this, we aim to enhance the robustness of detecting neovascular activity evolution.

The evaluation metrics for our best submission are shown in Table 1.

---

[2] https://dl.fbaipublicfiles.com/dinov2/dinov2_vitb14/dinov2_vitb14_reg4_pretrain.pth

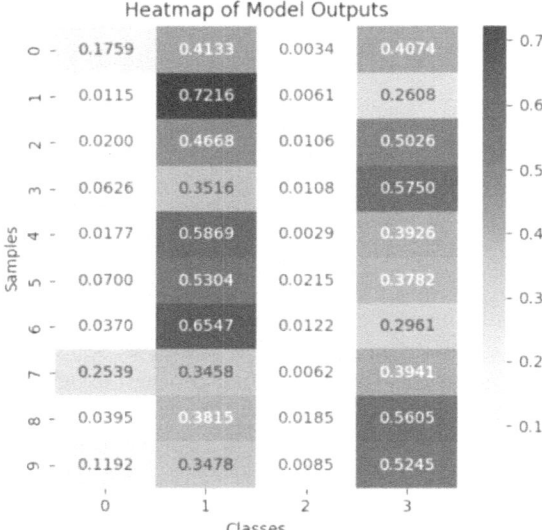

Fig. 2. Heatmap of model outputs for different samples and classes. One sample corresponds to one patient. Dataset details are provided in Sect. 2.1.

3 Results

Our best submission employs a DinoV2 ViT with a 10-fold cross-validation ensemble approach. We chose this approach based on performance on the public leaderboard. Stratification is applied using random seeds from 0 to 9. We train models for 10 epochs for each fold, selecting the checkpoint with the lowest validation loss for optimal performance. Surprisingly, the lowest loss was typically achieved within the first one or two epochs.

Overall, the results indicate that our ensemble approach and post-processing method result in robustly strong performance across the train, validation, and leaderboard datasets.

4 Discussion

The experimental results suggest that our method can effectively detect changes in neovascular activity, which is crucial for tailoring anti-VEGF treatments to individual patients.

Figure 2 shows the distribution of our final model's outputs over the different classes on the training split. Across all stratification splits, class 2 (Worsened) is hardly ever predicted. However, classes 0, 1, and 3 vary much more across seeds, showing that the stratification split vehemently influences the model's predictions and provides strong motivation for our post-processing method that leverages predictions from all k-fold trained models. The improved robustness of our models is crucial in such a critical task as ophthalmological image analysis.

One limitation of our method concerns the global feature extraction we employ. We extract features on resized images without regard to their actual position in the image and derive predictions from these features. Thus, we may lose out on fine-grained local information. This limitation in our current model presents an opportunity for future refinement, e.g., by segmenting images into different regions. Specifically, this could involve segmenting the images into different areas (e.g., A, B, C) and then comparing the pixel quantities of the segmented images to determine the classification [4,5]. However, implementing this approach would require a robust, generalizable preprocessing pipeline capable of flattening the images, aligning them, and adjusting them to the correct size. In practice, while promising, we could not find such a robust and generalizable pipeline. We leave this to future work. Specifically, we encountered significant challenges in implementing this segmentation approach, as the U-Net [10] we employed for segmentation did not generalize well to the challenge dataset, with noisy images being particularly problematic. We explored denoising techniques but could not implement an effective solution within the project timeframe. Additionally, finding an appropriate open dataset to train our segmentation U-Net proved challenging, although we identified a potential source [4]. However, we could not successfully integrate it into our final solution. For details and reference for future work, we refer to Appendix A.1.

Another approach we explored involved a data pre-processing step to generate new labels based on patient visits, as detailed in Appendix A.2. However, this method did not yield improved results. Instead, it decreased performance, which could indicate significant noise in the labels (possibly due to incorrect or ambiguous annotations).

These challenges and attempted solutions highlight the complexity of developing robust models for detecting changes in neovascular activity. While our current approach shows promise, there are clear avenues for improvement in future work. Moreover, while our method shows promise on this dataset, further validation on larger datasets and in diverse clinical settings remains necessary to confirm the generalizability of our approach.

In conclusion, our simple but effective model demonstrates strong performance in detecting changes in neovascular activity. Yet, there is potential for enhancement. This, however, has proved not to be straightforward. The challenges we encountered and the alternative approaches we explored provide valuable insights for future research directions in this critical area of ophthalmological image analysis.

5 Link to Public Code Repository

The implementation of our method, along with the pre-trained models and evaluation scripts, is available at the following public repository:

https://github.com/marceljhuber/mario-miccai2024

Acknowledgments. This research was supported by the Institute for Machine Learning of the Johannes Kepler University Linz. We are specifically grateful to Andreas Mayr and Niklas Schmidinger for their continued feedback and support. Furthermore, we are grateful to Hannah Aster, Petra Jósár, and Martin Marinschek for their continuous involvement in and contributions to the project.

Disclosure of Interests. The authors declare no competing interests.

A Additional Methods

A.1 Segmentation Approach

Here, we describe the segmentation approach we considered in Sect. 4. The approach involved segmenting the images into ten distinct areas to compare the pixel counts of these segmented regions for classification. The segmentation was performed using a U-Net architecture designed to segment the images into semantically meaningful regions [10].

An essential requirement for this method is a robust preprocessing pipeline to ensure accurate alignment and sizing of the images. This step is crucial to ensure that the segmented regions corresponded correctly between consecutive scans. Unfortunately, our U-Net model did not generalize well to the challenge dataset, primarily due to high noise levels in the images. Despite attempts to implement denoising techniques and integrate external datasets, the segmentation approach did not yield the desired improvements in performance.

A.2 Generation of New Labels

We outline our approach for generating new labels based on visit data to augment the training dataset, as mentioned in Sect. 4. Specifically, we generate new labels as a function of the existing sequence, resulting in up to $\frac{1}{2}(k-1)k$ new labels, where k is the sequence length. A sequence is defined as the set of visits filtered by patient ID, eye side, and identical B-Scan number, resulting in k data entries. By leveraging the current and subsequent visit IDs, we construct a label sequence that captures disease progression in a specific eye across k visits. This process is also described in Figure 3.

Recall the four distinct classes: *Reduced* (0), *Stable* (1), *Worsened* (2), and *Other* (3). Consider a sequence of $k = 9$ labels:

$$\{1 \to 1 \to 1 \to 1 \to 2 \to 1 \to 1 \to 2 \to 0\}$$

Translated into natural language, this sequence becomes:

{stable → stable → stable → stable → increased → stable → stable → increased → reduced}

Now, we assign values $s_i \in [0, 1]$ to these data points, where 0 indicates "healthy" and 1 represents "maximally diseased." These values are hypothetical

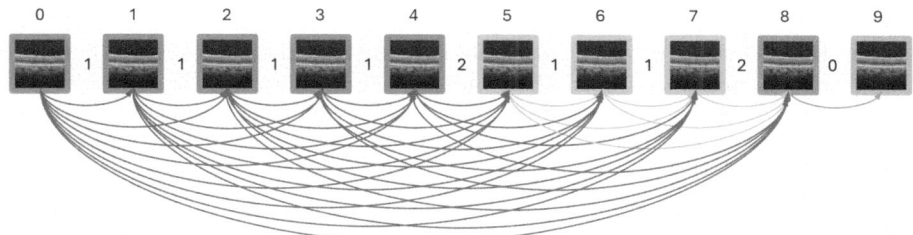

Fig. 3. Demonstration of new label generation. Each arrow between images represents a label. By default, arrows only exist between consecutive images. The color indicates clusters of images with the same level of disease severity (s_i). The numbers on top are the image indices, and the numbers between the images are the original labels from the tabular dataset. (Color figure online)

but reflect the progression of disease severity according to the labels. Also, note that while we have k labels, we have $k+1$ images, as each label compares two consecutive images. Thus, for $k+1=10$ images, we have:

$$\{\text{img}_1 \to \text{img}_2 \to \text{img}_3 \to \text{img}_4 \to \text{img}_5 \to \text{img}_6 \to \text{img}_7 \to \text{img}_8 \to \text{img}_9 \to \text{img}_{10}\}$$

Assigning exemplary values to these images based on the sequence of labels, we get the following:

$$\{0.25 \to 0.25 \to 0.25 \to 0.25 \to 0.25 \to 0.50 \to 0.50 \to 0.50 \to 0.75 \to 0.33\}$$

These values suggest that it is possible to cluster the images in our sequence based on the severity of disease s_i. We can directly compare clusters using the original labels by clustering the images. For example, consider clusters with severity values $s_i = 0.25$ and $s_i = 0.5$. We end up with two clusters of sizes five and three, respectively. We can then compare every image in cluster A with each in cluster B, resulting in fifteen labels. Initially, we only had a single label where the clusters intersected between images four and five.

Additionally, we can infer knowledge from images within the same cluster. For cluster A with $n = 5$ images, we generate $\sum_{i=1}^{n-1} i = 10$ labels. For cluster B with $n = 3$ images, we generate $\sum_{i=1}^{n-1} i = 3$ labels. Originally, we had only $4 + 2 = 6$ labels for these images. Using our proposed method, we obtain a total of $15 + 10 + 3 = 28$ labels by utilizing just two clusters.

The main limitation of this method becomes visible when clusters of label 0 (reduced) and 2 (worsened) follow each other because, based on the tabular data, there is no way of finding out the severity of change between s_i and s_j. Based on this info, the patient's eye severity level could hypothetically go from 0.66 to 0.659999 or 0; we can not tell. Moreover, if there is an occurrence of label 3 (other), we have no way of continuing to generate new labels in this sequence.

However, this approach did not improve the model's performance but led to a deterioration in overall results. This could indicate that the new labels were noisy or inaccurately annotated or that the method for generating these labels had fundamental issues. Detailed examination of the newly generated labels revealed inconsistencies and efforts to address these issues did not result in performance gains.

References

1. Darcet, T., Oquab, M., Mairal, J., Bojanowski, P.: Vision transformers need registers. ArXiv **abs/2309.16588** (2023). https://api.semanticscholar.org/CorpusID: 263134283
2. Dosovitskiy, A., et al.: An image is worth 16 × 16 words: transformers for image recognition at scale. http://arxiv.org/pdf/2010.11929
3. Ghosh, S., Zhao, X., Alim, M., Brudno, M., Bhat, M.: Artificial intelligence applied to 'omics data in liver disease: towards a personalised approach for diagnosis, prognosis and treatment. Gut (2024). https://api.semanticscholar.org/CorpusID: 271927861
4. Melinščak, M., Radmilović, M., Vatavuk, Z., Lončarić, S.: Annotated retinal optical coherence tomography images (aroi) database for joint retinal layer and fluid segmentation. AutomatikaâĂŕ: časopis za automatiku, mjerenje, elektroniku, računarstvo i komunikacije **62**(3–4), 375–385 (2021)
5. Melinščak, M., Radmilović, M., Vatavuk, Z., Lončarić, S.: Aroi: annotated retinal oct images database | ieee conference publication | ieee xplore. In: 2021 44th International Convention on Information, Communication and Electronic Technology (MIPRO) (2021)
6. Moor, M., et al.: Foundation models for generalist medical artificial intelligence. Nature **616**, 259–265 (2023). https://api.semanticscholar.org/CorpusID:258083369
7. Oquab, M., et al.: Dinov2: learning robust visual features without supervision. ArXiv **abs/2304.07193** (2023). https://api.semanticscholar.org/CorpusID: 258170077
8. Oquab, M., et al.: Dinov2: learning robust visual features without supervision, 29 August 2024. https://github.com/facebookresearch/dinov2,urldate=22.08.2024
9. Qiu, J., et al.: Visionfm: a multi-modal multi-task vision foundation model for generalist ophthalmic artificial intelligence. ArXiv **abs/2310.04992** (2023). https://api.semanticscholar.org/CorpusID:263828921
10. Ronneberger, O., Fischer, P., Brox, T.: U-net: convolutional networks for biomedical image segmentation. ArXiv **abs/1505.04597** (2015). https://api.semanticscholar.org/CorpusID:3719281
11. Stahl, A.: The diagnosis and treatment of age-related macular degeneration. Deutsches Ärzteblatt (117), 513–520 (2020). https://www.ncbi.nlm.nih.gov/pmc/articles/PMC7588619/
12. Thomas, C., Mirza, R., Gill, M.: Age-related macular degeneration. Med. Clin. **105**, 473–491 (2021)
13. Wiggins, W.F., Tejani, A.S.: On the opportunities and risks of foundation models for natural language processing in radiology. Radiology. Artif. Intell. **4 4**, e220119 (2022). https://api.semanticscholar.org/CorpusID:250934677
14. Zhou, Y., et al..: A foundation model for generalizable disease detection from retinal images. Nature **622**, 156–163 (2023). https://api.semanticscholar.org/CorpusID: 264168236

Monitoring Age-Related Macular Degeneration Progression in Optical Coherence Tomography (MARIO), Task 2 - MICCAI Challenge 2024, *jkulinzstudents* Submission

Patrick Binder, Marcel Huber(✉), and Markus Frohmann

Johannes Kepler University Linz, Linz, Austria
marceljulianhuber@gmail.com

Abstract. Age-related Macular Degeneration (AMD) is the most common cause of severe vision loss in individuals over 50, primarily affecting central vision. In its early stages, AMD often presents no symptoms, making early detection and monitoring its progression essential. Managing AMD effectively requires close observation of neovascular activity, especially its response to anti-VEGF treatments. In this work, we present a predictive model that forecasts AMD progression 90 days in advance using a single Optical Coherence Tomography (OCT) B-scan. Our approach leverages advanced deep learning techniques and a novel latent matching model to improve the accuracy of disease state predictions and guide anti-VEGF treatment strategies.

Keywords: Age-related Macular Degeneration (AMD) · Optical Coherence Tomography (OCT) · Deep Learning

1 Introduction

Age-related macular degeneration (AMD) is a significant and irreversible cause of vision loss affecting millions of people worldwide, particularly those over the age of 50. AMD primarily affects the macula, the central part of the retina responsible for detailed vision, causing progressive impairment in activities such as reading, recognizing faces, and driving. In its early stages, AMD is often asymptomatic, highlighting the importance of early detection, which can be achieved by detecting the presence of drusen, small yellowish deposits that form under the retina [3]. Although there is no cure, recent advances in technology and treatment approaches have significantly improved the management of AMD. Three key developments are changing the way AMD is diagnosed and treated:

Optical Coherence Tomography (OCT) is an advanced imaging technique that has revolutionized how AMD is diagnosed and monitored. By providing high-resolution, cross-sectional images of the retina, OCT allows healthcare professionals to visualize intricate layers of the retina in detail. This ability is essential for detecting early signs of AMD and assessing the progression of the disease [2].

The introduction of anti-VEGF (vascular endothelial growth factor) treatments represents a breakthrough in the treatment of wet AMD by injecting drugs that target and inhibit VEGF, a protein that promotes the growth of abnormal blood vessels in the eye. By reducing the impact of these vessels, anti-VEGF treatments can significantly slow vision loss and, in some cases, restore lost vision. Regular monitoring with OCT is essential to assess the effectiveness of the treatment and to make any necessary adjustments [4,5].

New possibilities for the diagnosis and treatment of AMD have been identified by integrating machine learning into medical imaging. By analyzing OCT scans, ML algorithms can identify subtle patterns and abnormalities that may indicate AMD, leading to earlier detection. These algorithms can also help predict disease progression and response to treatment, enabling more personalized and effective management strategies for patients [1].

Recent advances in anti-VEGF treatments have made the management of AMD progression more feasible, but effective treatment planning relies on the timely detection of changes in neovascular activity. To address this issue, we propose an automated method for predicting progression from a single OCT B-scan. Leveraging medical foundation models and incorporating additional data, our approach predicts disease progression over the next 90 days, ultimately improving treatment strategies for AMD patients undergoing anti-VEGF therapy.

2 Method

Our approach uses the *foundation model* RetFound [5] to predict the progression of retinal disease over 90 days using OCT scans. Given only the scan from the initial time point t_0 as input, we aim to predict one of three possible progressions: *reduced* (0), *stable* (1) or *increased* (2). The method is structured in two steps: a latent matching step and a disease progression step. First, we predict the embedding of the OCT image t_{90} after 90 days in the latent matching step. Secondly, in the disease progression step, both embeddings are used to predict the type of the modeled disease progression. Both steps rely on multi-layer perceptrons (MLPs) to predict both the future embedding and progression type.

2.1 Dataset

Within the MARIO challenge, each team was provided with two datasets by the challenge organizers. The first dataset for task 1 contains 14,496 pairs of OCT B-scans, taken at different visits. Each sample is annotated with the progression of the disease over the timeframe between the two visits and is assigned one of four class labels corresponding to the disease state. For the second task, 3,823 samples were provided, each consisting of only one OCT B-scan. The label

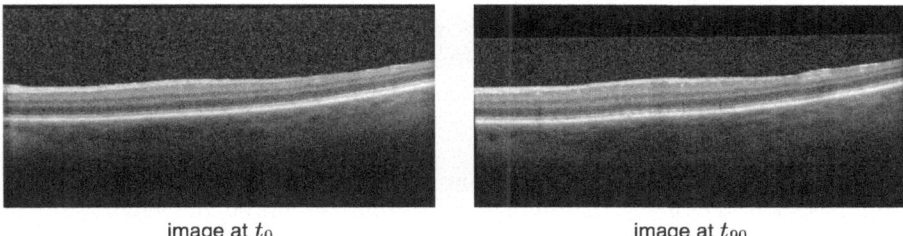

image at t_0 image at t_{90}

Fig. 1. Example from the training dataset of an OCT B-scan at t_0 (left) and t_{90} (right).

corresponding to each sample depicts how the condition progressed in 90 days as collected from consecutive medical examinations of 68 patients. Furthermore, both datasets provide additional information for each sample, like fundus images or data associated with the patient, such as their age and sex. For further details about the datasets, we refer to the challenge website.[1]

Since estimating disease progression utilizing only data from a single timestep is a challenging task, we solely rely on the dataset of task 1 but apply different pre-processing steps based on which step of our two-step approach. In the disease progression step, we apply the full available dataset of task 1.

For the latent matching step, we can only utilize a subset of the available data since, in this step, the disease progression over 90 days is learned to be estimated. Given an OCT B-scan taken at a visit at timestep t_a we thus filter for OCT B-scans taken at timestep t_b, where $t_b \in [t_a + 90 - n, t_a + 90 + n]$. This ensures that the time difference between the two visits deviates by not more than n days from the expected 90 days. This makes n a hyperparameter, trading off between the amount and quality of data for choosing a large and small n. In our experiments, we choose $n = 7$ as a condition fulfilled by 4,820 samples from the task 1 dataset. An example of two OCT B-scans with a time difference of 90 days is depicted in Fig. 1. When training the model for step 2, we apply a stratified split for the training and validation data, which closely resembles the performance on the public leaderboard and thus provides realistic estimates of the performance of our method.

2.2 Pre-processing

To enhance the performance of the individual models and our overall method, we apply a series of pre-processing steps that aim to shift the basis for the predictions to features relevant to neovascular activity. For inference, each OCT B-scan provided as input to our method is resized to a resolution of 224 × 224. This specific size is chosen since the feature extractor of our method relies on pre-trained images of this size, as further discussed in Sect. 2.4.

Since the latent matching step of our architecture builds directly on the feature vector given by our feature extraction, we also resize the inputs of this

---

[1] https://www.codabench.org/competitions/2852/.

step to 224 × 224 pixels and apply this also for step 2 of our architecture and normalize the data using the standard ImageNet statistics.

2.3 Proposed Method

To tackle the problem of predicting how the AMD condition will progress in a 90-day timeframe, given only information from the initial timepoint t_0, it is naturally a challenging task, as there is no information on how the condition progressed previously. Since a naive approach of predicting disease progression using information from t_0 only yielded unsatisfactory results, we draw inspiration from imagining the future state of the condition and making an informed prediction based on this information. We thus propose splitting the problem into two tasks—these are less complex and can be approached well utilizing the available data.

Specifically, we view the initial task as underconstrained since it requires predicting how the condition progresses in the next 90 days and whether this development can be seen as positive, negative, or neutral, given only information on the state at t_0. We thus transform the problem into two well-defined subproblems that are both approachable with the available data and require the respective models to learn a reasonable mapping. The first step aims to predict the state of the condition in 90 days, while the second step uses the initial data together with the prediction of step 1 to indicate the type of disease progression.

Latent Matching Step. The objective of this first step is to predict the state of the condition after 90 days, given the OCT B-scan at t_0. Given some ground truth B-scans from t_{90}, training an MLP on this mapping can solve this task well. Although this data is not available in the dataset for task 2, we extract a suitable subset from the one for task 1 as described in Sect. 2.1. Although this problem definition is already much simpler and more approachable than the initial problem, we still deem predicting the OCT B-scan at t_{90} a too challenging task, particularly given the limited amount of suitable training data. Therefore, we approach this task by finding a suitable mapping in the latent space of our applied feature extractor, pointing in the direction of a 90-day disease progression. As an input, we thus provide the extractor features to our model and train it on minimizing the negative cosine similarity between the predicted and ground truth embedding of the B-scan at t_90. We apply cosine similarity as a metric as it captures the similarity between two vectors while being normalized. The latent matching step is depicted in Fig. 2.

Disease Progression Step. In the second step of our method, we make use of the results of step 1 and build on the predictions of the latent matching model. Given an OCT B-scan from timestep t_0, we calculate the respective embedding e_0 by applying our chosen feature extractor and obtain an estimate of e_{90}, the embedding at timestep 90, by using the latent matching model from step 1. By doing so, we have access to the embeddings at both t_0 and t_{90}, and the overall task can be solved by comparing the two embeddings and classifying the observed progression instead of requiring to estimate the state of the condition at the same

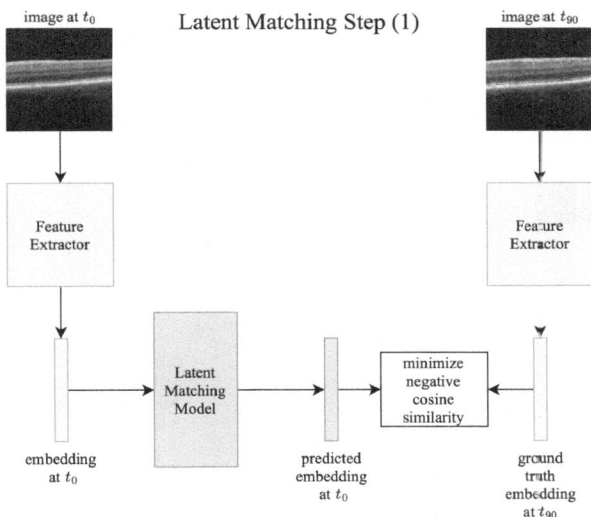

Fig. 2. Step (1): Latent Matching Step. The embeddings of the images at t_{90} are approximated given the embedding at timestep t_0, extracted from a frozen RetFound feature extractor.

time. This problem definition for the second step of our approach is thus also well suited to be solved by an MLP as a classification task. In particular, we concatenate the feature vectors e_0 and e_{90} and provide them jointly as input to the model, which predicts the corresponding class. Not only is this subproblem more approachable than the overall task, but the decomposition of task 2 renders the objective of the disease progression step the same way as to be solved in task 1. Therefore, we train the disease progression MLP on the full dataset available for task 1, minimizing cross entropy between the predicted and ground truth classes. The disease progression step is visualized in Fig. 3.

Ultimately, we thus divide the problem of predicting how the condition will change in 90 days, given information only on time step $t = 0$, into two simpler problems, with the second one being the problem to be solved in task 1.

Our latent matching model consists of 3 linear layers, with a constant number of 1024 neurons per layer, coupled with LeakyReLU as activation of the first two layers logits. The parameters are optimized using Adam with an initial learning rate of 1e-3, which is reduced by 0.9 each time the loss plateaus for 4 epochs. We train our method for 100 epochs and select the final model according to the performance on the validation set.

2.4 Pre-training

Due to limited computational resources, available data, and recent successes of foundation models, we employ the pre-trained RetFound model as a feature extractor. We discard the decoder layer since our approach only requires the

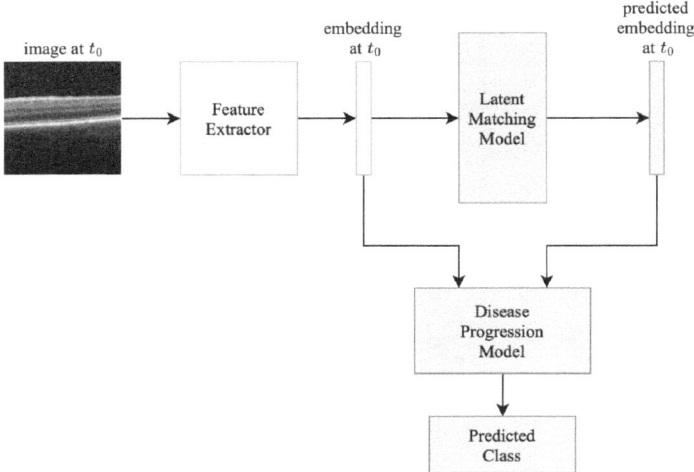

Fig. 3. Step (2): Disease Progression Step. The latent matching model of step 1 is being used to predict the embedding at t_{90} given the image at t_0. Both are fed into the disease progression model, which predicts the corresponding class.

encoder, the weight of which we freeze for efficiency. Fine-tuned on a large corpus of macular images, RetFound provides a solid basis for our approach and functions as a backbone for OCT images. Further details on RetFound can be found in [5].

We also experiment with fine-tuning RetFound on the provided OCT images. Specifically, we continue with the pre-training objective of masked image modeling, with a learning rate of 5e-5 for 10 epochs. Following [5], we discard the decoder and then only use the encoder for feature extraction. However, since we did not observe clear improvements when using this model's features on the challenge objective, we did not continue using it. We hypothesize that the original RetFound is already trained on highly similar OCT images, and we did not want to incur the risk of heavily overfitting on the training set.

2.5 Post-processing

While our approach does not require any sophisticated post-processing techniques, the fact that we train step 2 on the data of task 1 introduces a discrepancy that needs to be addressed. While task 2 differentiates between the 3 different classes *Reduced* (0), *Stable* (1) and *Increase* (2), task 1 further considers a fourth class *Uninterpretable* (3). Any predictions of class 3 thus have to be adapted to comply with the format of task 2. Due to the overrepresentation of class 1, we simply set any predictions of class 3 to class 1, as it is statistically the most likely. No further adjustments to the model predictions are performed.

3 Results

To evaluate our method, we report the metrics calculated on both the train data and our holdout validation set, as well as the scores provided on the leaderboard. We report the negative cosine similarity for the train and validation set of the latent matching model in Table 1. The left side of Table 2 contains the metrics calculated for step 2 of our proposed architecture, while the scores on the right side refer to the results of the overall pipeline as evaluated on the task 2 leaderboard.

Table 1. Performance metrics of the latent matching MLP.

Neg. Cos. Sim. ↓	Train	Validation
Latent matching MLP	−0.9868	−0.9849

Table 2. Performance Metrics. Left: Disease progression MLP, where the leaderboard metrics are taken from the leaderboard of task 1. Right: evaluation of the complete pipeline on the task 2 leaderboard.

Metric	Train	Validation	Leaderboard	Metric	Leaderboard
F1 Score ↑	0.8096	0.7339	0.72310	F1 Score ↑	0.68864
CE Loss ↓	0.4768	0.7542	–	Weighted Kappa ↑	0.07885
Matthew's Corr. ↑	0.7542	0.4088	0.41583	Matthew's Corr. ↑	0.07067
Specificity ↑	0.9046	0.8519	0.85345	Specificity ↑	0.68156

Our results on the train, validation, and leaderboard datasets indicate a strong performance for both steps 1 and 2 of our approach, as well as for the overall task, and support the results of our theoretical pre-analysis, claiming that decomposing the problem of task 2 into two well-defined subproblems leads to superior performance. However, the large discrepancy between the F1 score and Matthews correlation coefficient (MCC) can be attributed to the skewed dataset distribution. Our model from task 1 already struggles with this imbalance, and the issue is exacerbated in task 2 by using a similar training approach. This results in a compounding effect, as the model inherits the same problem of generalizing to underrepresented classes.

4 Discussion

In the presented approach, we propose to transform an underconstrained problem into two steps, each of which can be solved by an individual MLP using data that sufficiently supports the individual steps. We leverage a state-of-the-art foundation model and build on its predictions, rendering our approach both more efficient and powerful. Lastly, we show that while using the data from two consecutive visits for training, our proposed architecture generalizes the overall task of predicting disease progression well, given that data is from only a single timestep.

Although it is a reasonable approach from the perspective of the challenge, setting the predictions of class 3 by our disease progression model to class 1 should be approached with caution in practice. From a medical perspective, uninterpretable predictions should require special medical attention from professionals and not be considered stable, as this could potentially lead to false negative diagnoses. A more sophisticated approach could consider this and incorporate this class accordingly.

Further, the imbalanced class distribution can be considered to improve predictions on classes 0 and 2. Further, additional data from task 2 could be incorporated into our architecture, and the whole pipeline could be fine-tuned on predictions using this data. We leave this and other class imbalance mitigation strategies to future work.

Moreover, we select the training data using a $[-7, +7]$ window around the 90-day difference, as this provided a reasonable amount of samples while not diverging too far from the objective. Future work should thus consider the case of a narrower window, ideally using a larger dataset, leading to a higher number of high-quality samples.

5 Link to Public Code Repository

The implementation of our method, along with the pre-trained models and evaluation scripts, is available at the following public repository:

https://github.com/marceljhuber/mario-miccai2024

Acknowledgments. This research was supported by the Institute for Machine Learning of the Johannes Kepler University Linz. We are specifically grateful to Andreas Mayr and Niklas Schmidinger for their continued feedback and support. Furthermore, we are grateful to Hannah Aster, Petra Jósár, and Martin Marinschek for their continuous involvement in and contributions to the project.

Disclosure of Interests. The authors declare no competing interests.

References

1. Moor, M., et al.: Foundation models for generalist medical artificial intelligence. Nature **616**, 259–265 (2023). https://api.semanticscholar.org/CorpusID:258083369
2. Stahl, A.: The diagnosis and treatment of age-related macular degeneration. Deutsches Ärzteblatt (117), 513–520 (2020). https://www.ncbi.nlm.nih.gov/pmc/articles/PMC7588619/

3. Thomas, C., Mirza, R., Gill, M.: Age-related macular degeneration. Med. Clin. **105**, 473–491 (2021)
4. Wiggins, W.F., Tejani, A.S.: On the opportunities and risks of foundation models for natural language processing in radiology. Radiol. Artif. Intell. **4 4**, e220119 (2022). https://api.semanticscholar.org/CorpusID:250934677
5. Zhou, Y., et al.: A foundation model for generalizable disease detection from retinal images. Nature **622**, 156–163 (2023). https://api.semanticscholar.org/CorpusID:264168236

Modular Transformer-Based Monitoring and Prediction of Wet AMD Progression Using OCT Imaging

Lovre Antonio Budimir[(✉)][iD], Ivana Matovinović[iD], Donik Vršnak[iD], and Sven Lončarić[iD]

Faculty of Electrical Engineering and Computing, University of Zagreb, Unska 3, Zagreb, Croatia
{lovre-antonio.budimir,ivana.matovinovic,donik.vrsnak, sven.loncaric}@fer.unizg.hr

Abstract. Age-related macular degeneration (AMD) is a globally recognized age-related eye disease that can lead to severe visual impairment.This paper presents a novel application of modern computer vision techniques to monitor and predict the progression of wet AMD using Optical Coherence Tomography (OCT) imaging. The research in our work is structured around two tasks. The first task involves analyzing data from consecutive follow-up exams of a patient with wet AMD. In order to classify the evolution of neovascular activity between pairs of 2D OCT slices from these exams, a novel, fully transformer-based, three-part method is used to extract features from OCT B-scans and accurately detect and quantify changes in disease progression. The second task focuses on predicting neovascular activity within three months, using data from the current OCT exam. The goal is to develop a predictive model that can indicate the progression of wet AMD, enabling more informed and timely treatment planning. For the second task, we transfer knowledge from the key elements of the first task, which is considered simpler, into a two-part transformer-based method for extracting features and predicting disease evolution. We validate our methods on the MARIO-MICCAI 2024 Challenge dataset. The code is available at https://github.com/LovreAB17/FERLIV-MARIO.

Keywords: Transformer · Deep Learning · OCT · Wet AMD

1 Introduction

Age-related macular degeneration (AMD) has become the leading cause of vision loss in people over 60 in the developed world [23]. There are two AMD forms: atrophic or nonexudative (dry AMD) and neovascular or exudative (wet AMD). Dry AMD is manifested by thinning of the macula and gradually progresses through three stages: early, intermediate, and geographic atrophy (GA). On the other hand, wet AMD causes faster vision loss due to pathological blood vessels growing beneath or into the retina, known as choroid neovascularization. It causes leakage of fluid and blood, which subsequently damages the macula [4,11]. Unlike dry AMD, wet AMD can be successfully treated with monthly intravitreal injections of anti-vascular endothelial growth factors (anti-VEGFs) [17]. Delays in detection and inadequate treatment plans hamper the effectiveness of this method [18]. Furthermore, each anti-VEGF injection carries a low risk of various ocular complications, so alternative regimens with less frequent dosing are also being explored [10].

Monitoring AMD progress relies on optical coherence tomography (OCT), which allows cross-sectional visualization of retinal structures and disease-related features [7,21]. OCT and other imaging modalities generate large datasets that AI tools could utilize. In the context of wet AMD, developing an AI-based model that suggests the optimal anti-VEGF treatment would be highly beneficial [18]. Although there is progress in AMD research [2,8,24], it is still insufficient, given the disease's comprehensive health and social impact. Advancing the research in this area and improving the planning of anti-VEGF treatments was the primary motivation behind Monitoring Age-related Macular Degeneration Progression In Optical Coherence Tomography (MARIO) - MICCAI Challenge 2024 [1].

MARIO-MICCAI Challenge 2024 [1] tackles the following two tasks:

1. Monitoring the evolution of neovascular activity between two consecutive follow-up examinations of a patient with wet AMD.
2. Prediction of neovascular activity within three months using single exam data of a patient with wet AMD.

In our work, inspired by the achievements of vision transformers in computer vision [9,22] and retina image processing [25], we address these two tasks using novel transformer-based methods for monitoring and predicting wet AMD progression using OCT imaging.

For the first task, we develop a three-part method for classifying neovascular activity between pairs of OCT B-scans. The method starts with feature extraction from OCT B-scans, followed by a transformer-based Change Encoder that detects changes in retinal structures between follow-up exams. In the final part of the method, a transformer-based Diagnosis Encoder is used to process the changes and diagnose the progression of wet AMD. Subtle changes in the structures of some retinal layers, such as RPE, are critical in analyzing the wet AMD progression [15]. For this reason, the Feature Encoder is pre-trained on a retinal segmentation task. To further adapt to Task 1, the variability caused by inter-device transfer is addressed by using the public OCT dataset [6] acquired on the same device as the dataset provided by the MARIO Challenge 2024 [1].

The second task is more challenging than the first because we are trying to predict the progression of wet AMD in the near future. For this reason, we transfer knowledge of the Feature Encoder and Diagnosis Encoder learned in the first task to create a two-part method for processing a single OCT B-scan image. The modular implementation of our method for the first task allowed us to perform transfer learning and easily implement the method for the second task by removing the Change Encoder.

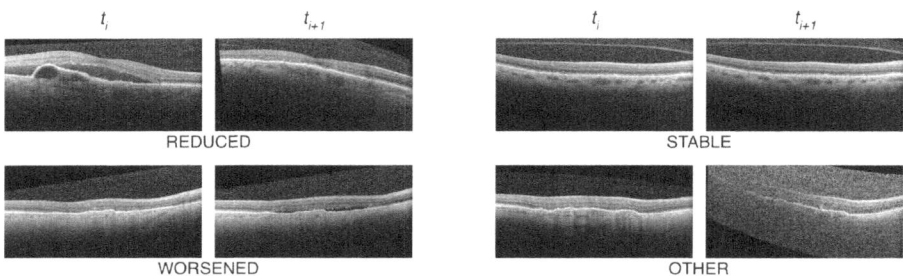

Fig. 1. Task 1 examples from MARIO-MICCAI Challenge dataset [1].

2 Dataset

OCT images of patients with wet AMD were acquired at the Ophthalmology department of Brest University Hospital, France, using a Spectralis OCT device. Two OCT C-scans from consecutive examinations for each patient were registered. The original B-scan resolution is 496 × 1024.

In Task 1, each case contains image data: a matched B-scan pair from two consecutive C-scans and the corresponding 2D infrared (IR) eye image pair, as well as patient-related data including eye side, sex, age, and the time difference between the two observed exams. Four disjunctive classes of vascular activity evolution between corresponding cases from two exams are provided: reduced, stable, worsened, and other, as shown in Fig. 1. The 'other' class refers to cases that could not be interpreted by a retina specialist or where the activity has disappeared.

In Task 2, each case contains a single B-scan with an associated 2D IR eye image and patient-related attributes such as eye side, sex, age, and the current exam number. A case should be categorized into one of three prediction classes: reduced, stable, or increased. Each class represents the extent of the evolution of vascular activity within three months after B-scan acquisition.

The dataset is highly imbalanced in both tasks. Table 1 contains both patient demographics and OCT volume statistics data for both tasks. The dataset is restricted only to registered challenge participants at the time of writing.

Table 1. Data statistics for Task 1 and Task 2.

	Task 1		Task 2	
	Train	Off-site validation	Train	Off-site validation
Patients	68	34	61	29
Female/Male	35/33	23/11	34/27	21/8
Age (mean/std)	81.22/8.54	80.68/8.53	83.20/6.30	82.79/7.24
OCT volumes	573	278	330	163
Number of cases	14496	7010	8082	3822

3 Method

3.1 Preprocessing

The data augmentation methods for training data are random cropping retaining 20–100% of the entire image, resizing to 224 × 224, random horizontal flipping, and image normalization using ImageNet [3] reference values. The data augmentation methods for testing data are resizing to 256 × 256, central crop to obtain the image size of 224 × 224, and image normalization using ImageNet reference values.

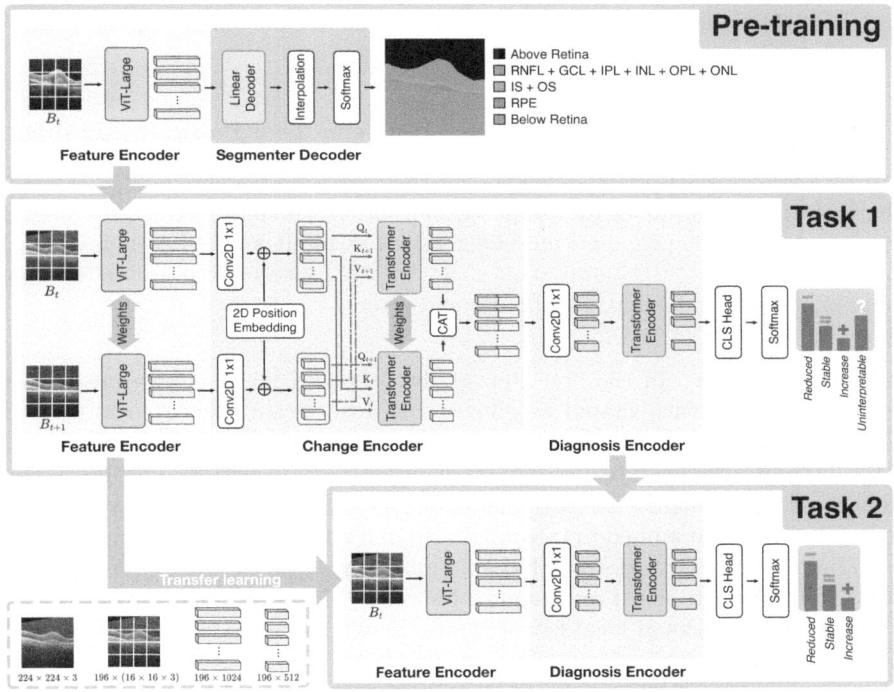

Fig. 2. Overview of our pre-training procedure and methods for Task 1 and Task 2.

3.2 Pretraining

The simpler version of Segmenter [20], with a point-wise linear decoder, as shown in Fig. 2 (Pre-training), was trained on the public Spectralis OCT dataset [6], which contains healthy and multiple-sclerosis OCT B-scans with annotated retinal layers. The model segments the vitreous and choroid as background regions, along with three retinal regions: the region between RNFL and ONL boundary, the IS/OS region, and the retinal pigment epithelium (RPE). The corresponding Feature Encoder in the pre-training segmentation task was initialized with OCT RETFound weights [25]. OCT RETFound is a large Vision Transformer (ViT-Large) [9], taken from the encoder part of the Masked Autoencoder [5], trained on 736,442 OCT images.

Training Details. We train our segmentation network using AdamW [12] for 30 epochs with $\beta_1 = 0.9$, $\beta_2 = 0.999$, weight decay 0.001, and a batch size of 8. We use a linear learning warm-up for five epochs with a starting learning rate of $1e - 6$. For the next 25 epochs, we use cosine learning rate decay [13] with a starting learning rate of $1e - 4$. The loss function is a cross-entropy loss. The model weights from the last epoch are used as the initial weights of the Feature Encoder of the following task.

3.3 Task 1: Classify the Evolution Between Two Pairs of 2-D Slices from Two Consecutive 2D OCT Acquisitions

The overall architecture of the proposed method for the first task is presented in Fig. 2 (Task 1). The proposed method receives as input a matched pair of OCT B-scans, $B_t \in \mathbb{R}^{224 \times 224 \times 3}$ and $B_{t+1} \in \mathbb{R}^{224 \times 224 \times 3}$, where t represents the time index of the medical exam, i.e., before and after.

Feature Encoder. In the first step of the proposed method, two pre-trained feature extractors with shared weights are used to obtain a set of local features from OCT B-scans. In our method, we follow the work of [25] by splitting the 224×224 input image into 196 non-overlapping 16×16 patches and projecting them into 1024-dimensional patch embeddings. We add the corresponding 2D sine-cosine position embedding to all 196 patch embeddings and forward them to ViT-Large's 24 Transformer blocks. Implementation details of Transformer blocks follow the standard implementation of ViT-Large [9]. The output is a set of 196 1024-dimensional local features for a single B-scan.

Change Encoder. The extracted local feature vectors from a pair of OCT B-scans are submitted to the second part of the method, which must encode visual changes in a retina between two medical examinations. For encoding changes between two medical exams, we use dual Multi-Change Captioning Transformer (MCCFormers-D) [16]. In our method, we follow [16] by using 1×1 convolution to project local features into 512-dimensional embeddings and add corresponding 2D learnable position embedding to each of the 196 local features. We utilize the co-attention mechanism [14] to capture the relationship between the local features of two B-scans. MCCFormers-D [16] uses two Transformer encoders with shared weights. The Multi-Head Attention (MHA) [22] of the first encoder is computed using queries from the first medical exam Q_t with keys K_{t+1} and values V_{t+1} from the second medical exam, and vice versa [16]. MCCFormers-D has two Transformer encoder blocks in our work, and the number of heads in MHA is 8. The dimensionality of the input and the output of the Transformer block d_{model} is 512, while the dimensionality of the inner-linear layers d_{ff} is 2048. The two encoders' corresponding outputs are concatenated, resulting in a new set of 196 1024-dimensional local change features.

Diagnosis Encoder. In the final part of our method, concatenated local features are projected once again using 1×1 convolution into 512-dimensional embeddings and forwarded to a new Transformer encoder without positional embeddings. The number of encoder blocks is 2, the number of heads in MHA is 8, d_{model} is 512, and d_{ff} is 2048. The outputs of the Transformer encoder are forwarded to a classification head (CLS Head). CLS Head uses maximum pooling and layer normalization to obtain a single 512-dimensional feature vector. Finally, a fully connected layer produces the output corresponding to the number of classes.

Training Details. We train our method for the first task using AdamW [12] for 25 epochs with $\beta_1 = 0.9$, $\beta_2 = 0.999$, weight decay 0.0005 for the Feature Encoder and 0.005 for the rest of the model, and a batch size of 32. We use a linear learning warm-up for five epochs with a starting learning rate of $1e-6$, $1e-5$, and $1e-5$ for the Feature Encoder, Change Encoder, and Diagnosis Encoder, respectively. For the next 20 epochs, we use cosine learning rate decay [13] with starting learning rates of $1e-5$, $1e-4$, and $1e-4$ for the Feature Encoder, Change Encoder, and Diagnosis Encoder, respectively. Ending learning rates are $1e-6$, $1e-5$, and $1e-5$ for the Feature Encoder, Change Encoder, and Diagnosis Encoder, respectively. We use a lower learning rate for feature extraction since the Feature Encoder is pre-trained on the segmentation task in Sect. 3.2. The loss function is a weighted categorical cross-entropy loss, where the weights are calculated using a number of samples in each class in the training dataset. Both B-scans use the same augmentation techniques from Sect. 3.1 to ensure they stay registered.

3.4 Task 2: Prediction of Evolution Within 3 Months of AMD on OCT 2D Slices for Planning Treatment Anti-VEGF

The second task is much more challenging than the first because it is necessary to predict the behaviour of the disease three months in advance using information from a single medical examination. Inspired by the information learned in the first task, which is considered simpler, our method for the second task is built using the feature extraction and Diagnosis Encoder using pre-trained weights from Task 1, as shown in Fig. 2 (Task 2). For this task, we only have one OCT B-scan $B_t \in \mathbb{R}^{224 \times 224 \times 3}$ used as input. The CLS Head is the only part of the model not pre-trained on the first task, given that the second task has three possible classes while the first task has four. All implementation details for the Feature Encoder and Diagnosis Encoder are listed in Sect. 3.3.

Table 2. Results for Task 1.

Pre-training	micro F1-score	Specificity	Rk-correlation	Mean metrics
ImageNet	78.531	90.064	58.230	75.608
OCT RETFound	81.769	**91.008**	63.656	78.811
Segmenter	**83.224**	90.678	**65.227**	**79.710**

Training Details. We train our method using AdamW [12] for 20 epochs with $\beta_1 = 0.9$, $\beta_2 = 0.999$, weight decay 0.005, and a batch size of 32. We use a linear learning warm-up for 5 epochs with a starting learning rate of $1e-5$. For the next 15 epochs, we use cosine learning rate decay [13] with starting learning rates of $1e-4$. To address the issue of class imbalance, we use a weighted categorical cross-entropy loss function and a weighted random sampler.

4 Results

Per the challenge organizer's request, the following evaluation metrics are used: micro F1-score, Specificity, and Rk-correlation coefficient. Additionally, in Task 2, the Quadratic-weighted Kappa is also included.

Results for Task 1 are presented in Table 2, where we compare the performance of our method using different initialization weights for Feature Encoder. Our pre-training segmentation task, described in Sect. 3.2, leads to improvements of +1.46% in F1-score and +1.57% in Rk-correlation. The confusion matrix for Task 1 is shown in Fig. 3 (left). Qualitative results are presented in Fig. 4, where we can see that our method can detect neovascular activity changes between two B-scans, where wet AMD symptoms are reduced or worsened.

The difficulty of Task 2 compared to Task 1 is evident in our results. On the 4th epoch of training, we achieved the best results with 60.00%, 69.38%, 8.98%, and 14.08% on F1-score, Specificity, Rk-correlation, and Quadratic-weighted Kappa, respectively. The confusion matrix for Task 2 is shown in Fig. 3 (right).

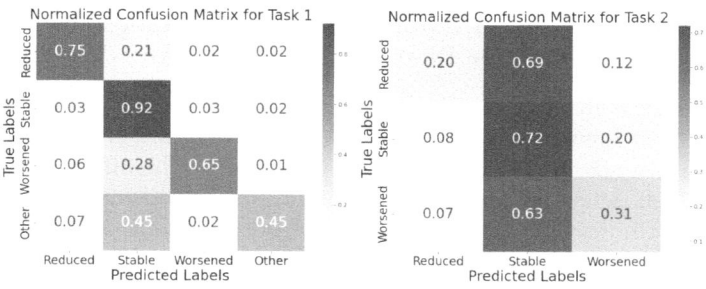

Fig. 3. Confusion matrix for Task 1 (left) and Task 2 (right).

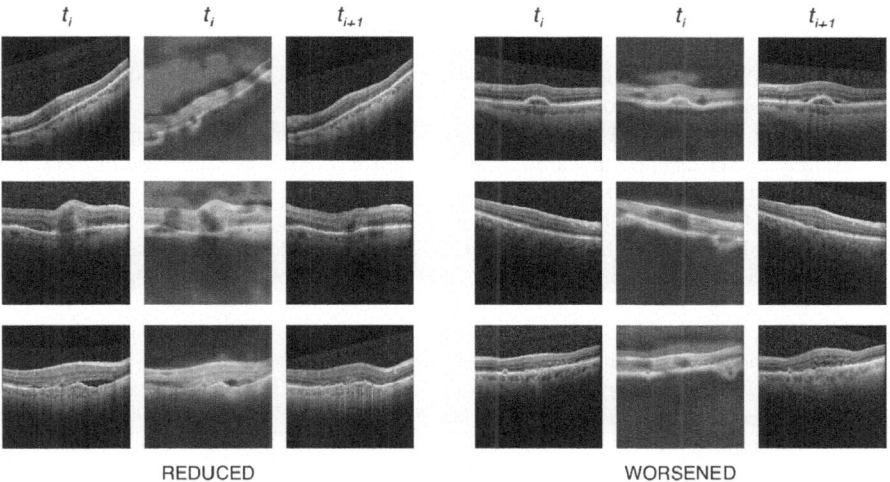

Fig. 4. Grad-CAM [19] visualization of examples with reduced and worsened cases on MARIO-MICCAI dataset [1].

5 Discussion

In this paper, we presented fully transformer-based methods for monitoring changes in neovascular activity between two OCT B-scans and predicting the progression of wet AMD within three months using a single OCT B-scan.

In Task 1, the preliminary results indicate that our three-part method for monitoring progression benefits from pre-training the Feature Encoder on the segmentation task. Although our method has demonstrated that it can identify areas where neovascular activity has occurred, problems are caused by a very unbalanced dataset dominated by many pairs of B-scans with no change between them, i.e., class "stable."

In Task 2, we transferred knowledge from Task 1 to Task 2, motivated by the benefits of transfer learning in the first task and the easy modularity of our method components. However, the high complexity of the second task, where we predict the future progression of wet AMD, combined with a very unbalanced dataset, led to much lower results compared to Task 1.

Our work fully relied on modern deep learning techniques, such as vision transformers and transfer learning, to create a competitive method for a complex computer vision problem. In future work, we will try to incorporate additional information from medical exams to create cross-modal methods and use preprocessing and postprocessing methods to improve our results on this issue further.

References

1. Mario challenge homepage. https://youvenz.github.io/MARIO_challenge.github.io/. Accessed 14 Sept 2024
2. Bogunović, H., et al.: Prediction of anti-VEGF treatment requirements in neovascular AMD using a machine learning approach. Invest. Ophthalmol. Vis. Sci. **58**(7), 3240–3248 (2017)
3. Deng, J., Dong, W., Socher, R., Li, L.J., Li, K., Fei-Fei, L.: ImageNet: a large-scale hierarchical image database. In: 2009 IEEE Conference on Computer Vision and Pattern Recognition, pp. 248–255 (2009). https://doi.org/10.1109/CVPR.2009.5206848
4. Deng, Y., et al.: Age-related macular degeneration: epidemiology, genetics, pathophysiology, diagnosis, and targeted therapy. Genes Dis. **9**(1), 62–79 (2022). https://doi.org/10.1016/j.gendis.2021.02.009. https://www.sciencedirect.com/science/article/pii/S2352304221000295
5. He, K., Chen, X., Xie, S., Li, Y., Dollár, P., Girshick, R.: Masked autoencoders are scalable vision learners. In: Proceedings of the IEEE/CVF Conference on Computer Vision and Pattern Recognition, pp. 16000–16009 (2022)
6. He, Y., Carass, A., Solomon, S.D., Saidha, S., Calabresi, P.A., Prince, J.L.: Retinal layer parcellation of optical coherence tomography images: data resource for multiple sclerosis and healthy controls. Data Brief **22**, 601–604 (2019)
7. Huang, D., et al.: Optical coherence tomography. Science **254**(5035), 1178–1181 (1991)
8. Jung, J., et al.: Prediction of neovascular age-related macular degeneration recurrence using optical coherence tomography images with a deep neural network. Sci. Rep. **14**(1), 5854 (2024)
9. Kolesnikov, A., et al.: An image is worth 16 × 16 words: transformers for image recognition at scale (2021)
10. Li, E., Donati, S., Lindsley, K.B., Krzystolik, M.G., Virgili, G.: Treatment regimens for administration of anti-vascular endothelial growth factor agents for neovascular age-related macular degeneration. Cochrane Database Syst. Rev. (5) (2020)
11. Lim, L.S., Mitchell, P., Seddon, J.M., Holz, F.G., Wong, T.Y.: Age-related macular degeneration. The Lancet **379**(9827), 1728–1738 (2012)
12. Loshchilov, I.: Decoupled weight decay regularization. arXiv preprint arXiv:1711.05101 (2017)
13. Loshchilov, I., Hutter, F.: SGDR: stochastic gradient descent with warm restarts. In: International Conference on Learning Representations (2017). https://openreview.net/forum?id=Skq89Scxx
14. Lu, J., Batra, D., Parikh, D., Lee, S.: ViLBERT: pretraining task-agnostic visiolinguistic representations for vision-and-language tasks. In: Advances in Neural Information Processing Systems 32 (2019)
15. Melinščak, M., Radmilović, M., Vatavuk, Z., Lončarić, S.: Annotated retinal optical coherence tomography images (AROI) database for joint retinal layer and fluid segmentation. Automatika: časopis za automatiku, mjerenje, elektroniku, računarstvo i komunikacije **62**(3–4), 375–385 (2021)
16. Qiu, Y., et al.: Describing and localizing multiple changes with transformers. In: Proceedings of the IEEE/CVF International Conference on Computer Vision, pp. 1971–1980 (2021)
17. Rosenfeld, P.J., et al.: Ranibizumab for neovascular age-related macular degeneration. N. Engl. J. Med. **355**(14), 1419–1431 (2006)

18. Schmidt-Erfurth, U., Sadeghipour, A., Gerendas, B.S., Waldstein, S.M., Bogunović, H.: Artificial intelligence in retina. Prog. Retin. Eye Res. **67**, 1–29 (2018)
19. Selvaraju, R.R., Cogswell, M., Das, A., Vedantam, R., Parikh, D., Batra, D.: Grad-CAM: visual explanations from deep networks via gradient-based localization. In: Proceedings of the IEEE International Conference on Computer Vision, pp. 618–626 (2017)
20. Strudel, R., Garcia, R., Laptev, I., Schmid, C.: Segmenter: transformer for semantic segmentation. In: 2021 IEEE/CVF International Conference on Computer Vision (ICCV), pp. 7242–7252 (2021)
21. Swanson, E.A., et al.: In vivo retinal imaging by optical coherence tomography. Opt. Lett. **18**(21), 1864–1866 (1993)
22. Vaswani, A.: Attention is all you need. In: Advances in Neural Information Processing Systems (2017)
23. Wong, W.L., et al.: Global prevalence of age-related macular degeneration and disease burden projection for 2020 and 2040: a systematic review and meta-analysis. Lancet Glob. Health **2**(2), 106–116 (2014)
24. Yeh, T.C., et al.: Prediction of treatment outcome in neovascular age-related macular degeneration using a novel convolutional neural network. Sci. Rep. **12**(1), 5871 (2022)
25. Zhou, Y., et al.: A foundation model for generalizable disease detection from retinal images. Nature **622**(7981), 156–163 (2023)

A Novel Multimodal Deep Learning Fusion Framework for Predicting Neovascular Activity Evolution in Exudative Age-Related Macular Degeneration

Christopher Nielsen[1,2]([✉]), Ahmad O. Ahsan[1,2], Matthias Wilms[1,3,4,5,6], and Nils D. Forkert[1,3,4,7]

[1] Department of Radiology, University of Calgary, Calgary, AB, Canada
christopher.nielsen@ucalgary.ca
[2] Biomedical Engineering Graduate Program, University of Calgary, Calgary, AB, Canada
[3] Hotchkiss Brain Institute, University of Calgary, Calgary, AB, Canada
[4] Alberta Children's Hospital Research Institute, University of Calgary, Calgary, AB, Canada
[5] Department of Pediatrics, University of Calgary, Calgary, AB, Canada
[6] Department of Community Health Sciences, University of Calgary, Calgary, AB, Canada
[7] Department of Clinical Neuroscience, University of Calgary, Calgary, AB, Canada

Abstract. Age-related macular degeneration (AMD) is a leading cause of vision loss among older adults, characterized by the progressive deterioration of the macula. Accurate monitoring and prediction of AMD progression is crucial for effective treatment planning and improved patient outcomes. This work presents a novel multimodal fusion deep learning framework designed to automate the analysis of optical coherence tomography (OCT) images for two primary tasks defined in the MICCAI 2024 Monitoring AMD Progression in OCT (MARIO) challenge: (1) classifying changes between 2D OCT B-scans captured at consecutive time points, and (2) predicting the three-month evolution of retinal structure based on OCT data from a single time point. The developed framework leverages finetuned RETFound and EfficientNetV2 models to extract feature representations from OCT B-scans and infrared fundus localizer images, while also incorporating additional clinical variables to enhance prediction accuracy. Extensive experiments conducted on the MARIO challenge dataset demonstrate the effectiveness of the proposed framework, with the best models achieving F1 scores of 0.851 for Task 1 and 0.703 for Task 2. These findings underscore the potential of integrating multimodal data to facilitate an automated analysis and prediction of AMD progression, paving the way for more effective and personalized treatment strategies. Software for the developed framework is available at https://github.com/chrisnielsen/miccai-2024-mario-challenge.

Keywords: Age-related macular degeneration · optical coherence tomography · disease progression prediction · machine learning

1 Introduction

Age-related macular degeneration (AMD) is a leading cause of visual impairment and irreversible blindness among the elderly population, affecting approximately 200 million individuals globally [1]. AMD is typically classified into two forms: dry (non-exudative) and wet (exudative) [2]. Dry AMD, characterized by the gradual thinning of the macula and the accumulation of drusen, is the more common and less severe form of the disease. However, it can progress to the more damaging wet AMD, which involves the growth of abnormal blood vessels under the retina and the macula. These vessels are prone to leakage, leading to swelling, bleeding, and rapid and severe loss of central vision [3].

The pathogenesis of AMD is influenced by a combination of genetic, environmental, and metabolic factors [4]. Despite significant advances in exploring the molecular and cellular processes involved, the exact mechanisms driving AMD progression remain complex and multifaceted. Advances in imaging techniques, particularly optical coherence tomography (OCT), have revolutionized the diagnosis and monitoring of AMD [5]. OCT provides high-resolution, cross-sectional images of the retina, allowing for the precise assessment of the structural changes associated with the progression of AMD. Early and accurate detection of changes indicative of progression from dry to wet AMD is crucial for timely intervention and can significantly alter the therapeutic outcome.

The introduction of anti-vascular endothelial growth factor (anti-VEGF) treatments has significantly altered the therapeutic landscape for wet AMD by inhibiting the proliferation of abnormal blood vessels beneath the retina [6]. Despite these advances, the management of AMD with anti-VEGF therapy requires meticulous monitoring to tailor treatment plans effectively [7]. This monitoring is largely dependent on various biomarkers detectable through OCT imaging [8]. Biomarkers such as retinal thickness, the presence of subretinal or intraretinal fluid, and the integrity of the retinal pigment epithelium are critical indicators typically used to assess the activity of neovascular lesions. Tracking these biomarkers over time allows clinicians to evaluate the response to anti-VEGF treatment and adjust dosing intervals and treatment strategies accordingly. Specifically, observing the evolution of neovascular activity in OCT scans is essential for determining the frequency and dosage of anti-VEGF injections, potentially leading to more personalized and effective treatment regimens [9].

Amidst these developments, artificial intelligence (AI) systems have emerged as a promising tool to enhance the precision and efficiency of AMD management [10]. AI technologies, particularly machine learning models trained on large datasets of OCT images, can potentially further automate and support the detection and quantification of imaging biomarkers, thus, providing faster and possibly more accurate assessments of AMD disease progression and treatment response. The integration of AI into clinical practice not only holds great promise to improve patient outcomes by optimizing treatment plans but also has the potential to revolutionize the monitoring processes by enabling more dynamic and responsive AMD management [11].

Therefore, the purpose of this work is to develop a novel multimodal fusion framework to automate the analysis of OCT images for two critical clinical tasks as defined in the MICCAI 2024 Monitoring AMD Progression in OCT (MARIO) challenge [12]. Specifically, Task 1 focuses on comparing methods for classifying changes between

2D OCT B-scans acquired at two consecutive time points, while Task 2 aims to compare novel methods for predicting whether AMD progression will decrease, stabilize, or worsen over a three-month period from OCT data captured at a single time point. In this work, we finetune and utilize RETFound [13], a powerful foundation model for retinal image analysis, in combination with an EfficientNetV2 [14] convolutional neural network (CNN) architecture to extract feature representations from OCT B-scans and infrared fundus localizer images. These extracted image features, combined with other recorded clinical variables (*e.g.*, patient age and sex), are then used to train multimodal fusion prediction models for both tasks.

The main contributions of this work are: 1) Developing a novel multimodal fusion model to enable a highly accurate computer-aided AMD progression analysis and prediction, and 2) Conducting extensive experiments on the MARIO challenge dataset to evaluate the performance of the proposed framework for both tasks.

2 Material and Methods

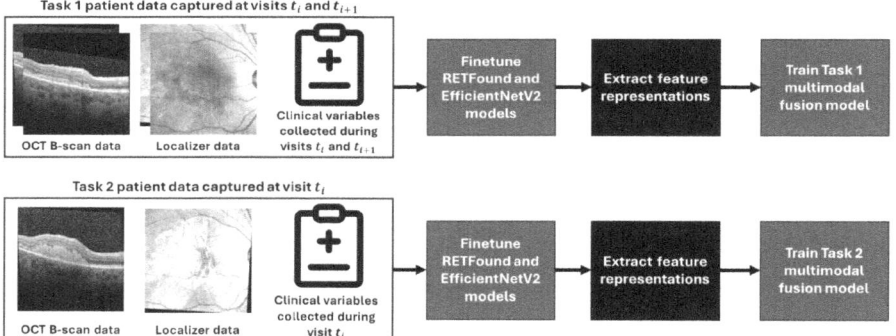

Fig. 1. Workflow for the developed multimodal fusion framework.

Figure 1 provides a high-level overview of the developed multimodal fusion framework. Initially, patient data comprising of 2D OCT B-scans (two B-scans captured at times t_i and t_{i+1} for Task 1, and a single B-scan captured at time t_i for Task 2), corresponding infrared localizer images, and clinical variables are used to finetune the RETFound [13] and EfficientNetV2 [14] architectures. Once finetuning is complete, these models extract lower-dimensional feature representations from the OCT B-scan and localizer image data. The motivation for using both RETFound, a transformer-based architecture, and EfficientNetV2, a CNN-based architecture, for feature extraction across these imaging modalities is to combine the strength of RETFound in capturing global contextual information with the capability of EfficientNetV2 to process detailed spatial features. This complementary approach aims to enhance performance compared to relying on a single model architecture. These extracted features, combined with the patient's clinical variables, are then used to train the multimodal fusion prediction models for Task 1 (support vector classification model) and Task 2 (ordinal logistic classification model). The specific steps of this framework are explained in greater detail in the following sections.

2.1 Finetuning and Feature Representation Extraction

The RETFound model, which is publicly available, uses a masked auto-encoder framework consisting of encoder and decoder modules. The encoder is built on a large-scale vision transformer (ViT-large) architecture [15], which includes 24 transformer blocks designed to generate embedding vectors for each 16 × 16 pixel patch of an input image. Each transformer block features a multilayer perceptron and a multi-headed self-attention mechanism. The features obtained from these image patches are combined using global average pooling, producing representations with a dimensionality of 1,024 per input image. For this work, we utilized two RETFound encoders: one for OCT B-scan data and another one for infrared fundus localizer images. The input OCT B-scan slice data has a resolution of 496 × 496 pixels, while the input localizer images have a resolution of 384 × 384 pixels. Figure 2 provides an overview of the process for extracting feature representations with RETFound.

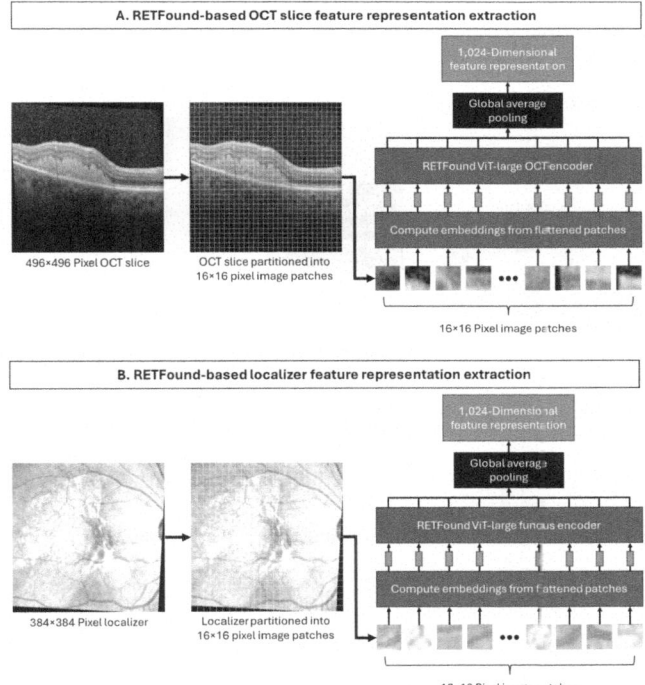

Fig. 2. RETFound feature representation extraction for: A) 2D OCT B-scan slices and B) infrared localizer fundus images.

EfficientNetV2 [14] is a CNN architecture, which includes inverted residual blocks with squeeze-and-excitation layers to capture channel-wise dependencies, along with a gradual increase in width to maintain a balance between model complexity and spatial detail. The EfficientNetV2 version used in this work contains about 20 million parameters and was pretrained on the ImageNet dataset [16] of natural images. Features extracted using EfficientNetV2 correspond to the activations of its penultimate layer, resulting in feature representations with a dimensionality of 1,796 for each input image. Two EfficientNetV2 models were employed to extract feature representations from OCT B-scan slice data (with an input resolution of 496 × 496 pixels) and localizer images (with an input resolution of 384 × 384 pixels).

To refine the feature representations generated by the RETFound and EfficientNetV2 models, we performed finetuning with a linear classification head for the Task 1 and Task 2 prediction tasks. The RETFound model was finetuned for 50 epochs, using a batch size of 1. Optimization was performed using AdamW [17] with a learning rate of 5×10^{-4}, and a weight decay of 0.05 to minimize cross-entropy loss. For EfficientNetV2, finetuning was conducted for 50 epochs, with a batch size of 6, utilizing the Adam optimizer [18] with a learning rate of 3×10^{-5} and a weight decay of 5×10^{-4} to minimize the cross-entropy loss. During the training phase for both models, data augmentation techniques, including random horizontal and vertical flips, rotation, translation, and contrast adjustments, were applied. All models were implemented in Pytorch version 1.13.1 and finetuning was performed using an Nvidia RTX 3090 GPU.

2.2 Multimodal Fusion Models

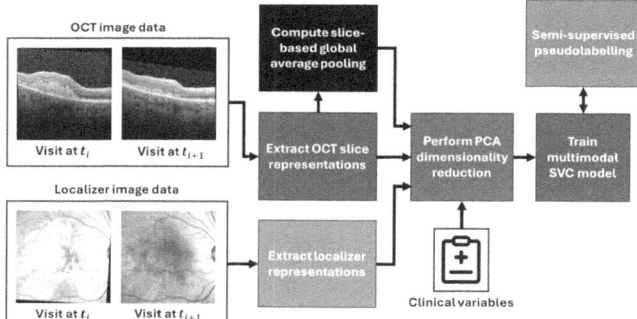

Fig. 3. Framework developed to train the multimodal support vector classification (SVC) model for Task 1.

For Task 1, which aims to classify AMD progression between two time points, the framework illustrated in Fig. 3 was developed. First, feature representations are extracted from all 2D OCT B-scan slices captured at times t_i and t_{i+1} using the finetuned RETFound and EfficientNetV2 models. To incorporate 3D contextual information when making predictions at the 2D B-scan slice level, the global average pooling of all 2D B-scan feature representations within each 3D C-scan image are also computed as additional features. Feature representations are also extracted from each 2D infrared localizer image captured at times t_i and t_{i+1} using the finetuned RETFound and EfficientNetV2 models. Clinical variables, including the patient's sex, age at each acquisition time point, visit number, and the time difference in days between acquisitions, were standardized using z-score normalization (subtracting the mean and dividing by the standard deviation), and these normalized variables were also included as features. Next, principal component analysis (PCA) is employed to reduce the dimensionality of all features to 200 to reduce the risk of overfitting. The resulting dimensionality-reduced principal components (PCs) are then used to train a support vector classification (SVC) model [19] with a hinge loss, a C-regularization coefficient of 1, and a radial basis function kernel with gamma set to the inverse of the number of features times the variance of the input data. After the initial training, a semi-supervised learning approach is employed to enhance the diversity of the training set and help regularize the model to prevent overfitting. This is achieved by assigning pseudolabels [20] to unlabeled data in the validation set, using high-confidence predictions with a probability greater than 0.8. The model is then retrained on a combined dataset consisting of the original training data and the pseudolabeled data. The SVC model is implemented and trained using scikit-learn 1.0.2 on an Intel i7-11700KF CPU.

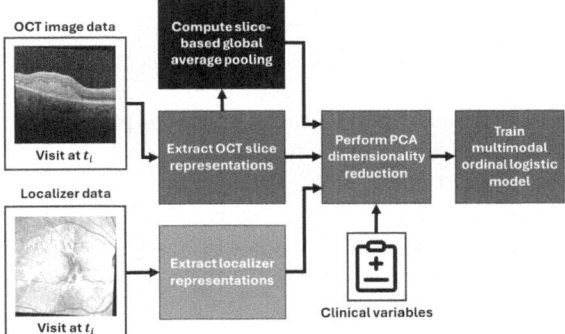

Fig. 4. Framework built to train multimodal ordinal logistic classification model for Task 2.

For Task 2, which aims to predict AMD progression over a three-month period, the framework presented in Fig. 4 was developed. Similar to Task 1, feature representations are extracted from each infrared localizer image and 2D OCT B-scan slices captured at time t_i. Likewise, to account for 3D contextual information when making predictions at the 2D B-scan slice level, global average pooling is performed on all 2D B-scan feature representations within each 3D C-scan image, adding these pooled features as additional inputs. Next, clinical variables, including the patient's sex, age at acquisition, and visit number, were standardized using z-score normalization and incorporated as features. PCA is then employed to reduce the dimensionality of all features to 200 to reduce the risk of overfitting. Unlike Task 1, Task 2 in the MARIO challenge uses the quadratic-weighted kappa score as a performance evaluation metric. To optimize performance for this metric, an ordinal logistic classification model is trained, employing the immediate-threshold loss variant described by Rennie et al. [21], with an L2 regularization coefficient of 0.05. The ordinal logistic classification model is implemented and trained using mord version 0.7 [22] on an Intel i7-11700KF CPU.

3 Experiments and Results

3.1 Dataset

All experiments in this work were conducted using only data from the MARIO challenge dataset [12]. The training dataset comprised 68 patients, while the validation dataset included 34 patients. For each patient, data encompassed OCT C-scan images from both the left and right eyes, captured at multiple time points. The imaging data was obtained using a Spectralis OCT device. The number of B-scans per C-scan varied by patient, ranging from 20 to 194. The resolution of each 2D B-scan was also patient-specific, with a fixed height of 496 pixels and a width ranging from 512 to 1536 pixels. In addition to OCT images, 2D infrared fundus images were collected to localize the spatial acquisition location for each 3D C-scan. All B-scan slices were resized to a resolution of 496 × 496 pixels, and all infrared fundus localizer images were resized to a resolution of 384 × 384 pixels during preprocessing. Clinical data, including patient sex and age at the time of imaging, was also gathered. To study disease progression, the MARIO challenge provides annotations that indicate whether AMD progression had decreased, stabilized, or worsened, focusing on two main tasks. Task 1 involves classifying changes between pairs of 2D B-scan slices from two consecutive OCT scans, similar to the "before" and "after" comparisons typically performed by clinicians [12]. Annotations for Task 1 also specify when B-scan slices are uninterpretable. For Task 1, 14,496 paired B-scans were annotated and used for training, with 7,010 paired B-scans reserved for validation. The second task centers on predicting the progression of 2D B-scan slices over a 3-months period, utilizing 8,082 annotated B-scans for training and 3,822 for validation. All annotations were provided by retina specialists with at least two years of experience in monitoring patients with vascular AMD.

3.2 Ablation Experiments and Results

To assess the effectiveness of the developed framework, ablation experiments were conducted to analyze the prediction performance for Task 1 and Task 2 when trained on different subsets of available input features. The performance results of the trained models on the validation dataset are presented in Table 1 for Task 1 and Table 2 for Task 2. The input feature subsets included: 1) feature representations from the 2D OCT B-scan slices (denoted in Tables as OCT-slice), 2) feature representations from the global average pooling of all 2D B-scan slices within each 3D C-scan (denoted in Tables as OCT-GAP), 3) feature representations from the localizer images (denoted in Tables as Localizer), and 4) tabular clinical patient variables (denoted in Tables as Clinical variables). For both tasks, the F1 score, R_k correlation coefficient, and specificity were measured. The performance of the deep learning model for Task 2 was also measured using the quadratic-weighted kappa (QWK) score as an additional metric. For Task 1, the model that demonstrated the best overall performance across all metrics was the one

Table 1. Performance metrics for the Task 1 ablation study. Each row presents the performance metrics calculated on the validation set for a model trained using a subset of the available features. OCT-GAP: global average pooled OCT features.

OCT-slice	OCT-GAP	Localizer	Clinical variables	F1 score	Specificity	Rk-correlation
✓	✗	✗	✗	0.846	0.897	0.670
✓	✗	✗	✓	0.846	0.898	0.670
✓	✗	✓	✗	0.848	0.900	0.674
✓	✗	✓	✓	0.849	0.901	0.676
✗	✓	✗	✗	0.720	0.796	0.311
✗	✓	✗	✓	0.722	0.796	0.315
✗	✓	✓	✗	0.720	0.800	0.316
✗	✓	✓	✓	0.720	0.800	0.316
✓	✓	✗	✗	0.849	0.900	0.678
✓	✓	✗	✓	0.850	0.900	0.679
✓	✓	✓	✗	0.850	0.901	0.680
✓	✓	✓	✓	**0.851**	**0.902**	**0.682**

trained on the complete set of features, achieving an F1 score of 0.851, a specificity of 0.902, and an R_k correlation coefficient of 0.682. In Task 2, the model with the best overall performance was trained exclusively on the OCT-GAP feature representations, achieving an F1 score of 0.703, specificity of 0.705, R_k correlation coefficient of 0.174, and a QWK score of 0.290.

Table 2. Performance metrics for the Task 2 ablation study. Each row presents the performance metrics calculated on the validation set for a model trained using a subset of the available features. OCT-GAP: global average pooled OCT features; QWK: quadratic-weighted kappa score.

OCT-slice	OCT-GAP	Localizer	Clinical variables	F1 score	Specificity	Rk-correlation	QWK
✓	✗	✗	✗	0.623	0.655	-0.025	0.038
✓	✗	✗	✓	0.611	0.651	-0.038	0.030
✓	✗	✓	✗	0.617	0.652	-0.033	0.055
✓	✗	✓	✓	0.619	0.654	-0.026	0.058
✗	✓	✗	✗	**0.703**	**0.705**	**0.174**	**0.290**
✗	✓	✗	✓	0.698	0.699	0.153	0.270
✗	✓	✓	✗	0.696	0.695	0.134	0.240
✗	✓	✓	✓	0.675	0.687	0.103	0.224
✓	✓	✗	✗	0.670	0.687	0.088	0.146
✓	✓	✗	✓	0.659	0.681	0.065	0.126
✓	✓	✓	✗	0.647	0.672	0.037	0.109
✓	✓	✓	✓	0.646	0.673	0.039	0.112

4 Discussion

In this work, we developed a multimodal fusion deep learning framework to predict AMD progression for two key tasks: 1) classifying changes using data from two consecutive exams (Task 1), and 2) predicting three-month progression based on data from a single exam (Task 2). Interestingly, the best-performing model for Task 1 utilized all available features during training, while the top model for Task 2 relied solely on global average pooling feature representations from the 2D OCT B-scan images. These results suggest that the prediction of three-month progression from a single timepoint benefits from incorporating data from all 2D B-scans within the 3D C-scan to improve accuracy. Future work will focus on developing an enhanced 2D B-scan aggregation strategy, replacing the equal weighting of slices in global average pooling with an attention mechanism to selectively integrate information from neighboring 2D B-scans. Additionally, class-balanced loss functions will be investigated to improve model performance and generalizability. In conclusion, this work has the potential to advance the computer-aided analysis of AMD progression, paving the way for more dynamic and tailored treatment strategies.

References

1. Vyawahare, H., Shinde, P.: Age-related macular degeneration: epidemiology, pathophysiology, diagnosis, and treatment. Cureus **14**(9), e29583 (2022). https://doi.org/10.7759/cureus.29583
2. Deng, Y., et al.: Age-related macular degeneration: epidemiology, genetics, pathophysiology, diagnosis, and targeted therapy. Genes Dis. **9**(1), 62–79 (2022). https://doi.org/10.1016/j.gendis.2021.02.009
3. Alexandru, M.R., Alexandra, N.M.: Wet age related macular degeneration management and follow-up. Rom. J. Ophthalmol. **60**(1), 9–13 (2016)
4. Wong, J.H.C., et al.: Exploring the pathogenesis of age-related macular degeneration: a review of the interplay between retinal pigment epithelium dysfunction and the innate immune system. Front. Neurosci. **16** (2022). https://doi.org/10.3389/fnins.2022.1009599
5. Elsharkawy, M., et al.: Role of optical coherence tomography imaging in predicting progression of age-related macular disease: a survey. Diagnostics **11**(12), 2313 (2021). https://doi.org/10.3390/diagnostics11122313
6. Song, D., Liu, P., Shang, K., Ma, Y.: Application and mechanism of anti-VEGF drugs in age-related macular degeneration. Front. Bioeng. Biotechnol. **10**, 943915 (2022). https://doi.org/10.3389/fbioe.2022.943915
7. Amoaku, W.M., et al.: Defining response to anti-VEGF therapies in neovascular AMD. Eye **29**(6), 721–731 (2015). https://doi.org/10.1038/eye.2015.48
8. Metrangolo, C., et al.: OCT biomarkers in neovascular age-related macular degeneration: a narrative review. J. Ophthalmol. **2021**, 9994098 (2021). https://doi.org/10.1155/2021/9994098
9. Moon, S., et al.: Prediction of anti-vascular endothelial growth factor agent-specific treatment outcomes in neovascular age-related macular degeneration using a generative adversarial network. Sci. Rep. **13**, 5639 (2023). https://doi.org/10.1038/s41598-023-32398-7
10. Dong, L., Yang, Q., Zhang, R.H., Wei, W.B.: Artificial intelligence for the detection of age-related macular degeneration in color fundus photographs: a systematic review and meta-analysis. eClinicalMedicine **35** (2021). https://doi.org/10.1016/j.eclinm.2021.100875
11. Crincoli, E., Sacconi, R., Querques, L., Querques, G.: Artificial intelligence in age-related macular degeneration: state of the art and recent updates. BMC Ophthalmol. **24**, 121 (2024). https://doi.org/10.1186/s12886-024-03381-1
12. Zeghlache, R., et al.: Monitoring age-related macular degeneration progression in optical coherence tomography (2024). https://doi.org/10.5281/zenodo.10992295
13. Zhou, Y., et al.: A foundation model for generalizable disease detection from retinal images. Nature **622**, 7981 (2023). https://doi.org/10.1038/s41586-023-06555-x

14. Tan, M., Le, Q.V.: EfficientNetV2: smaller models and faster training (2021). arXiv: arXiv: 2104.00298. https://doi.org/10.48550/arXiv.2104.00298
15. Dosovitskiy, A., et al.: An image is worth 16 × 16 words: transformers for image recognition at scale (2021). arXiv: arXiv:2010.11929. https://doi.org/10.48550/arXiv.2010.11929
16. Deng, J., Dong, W., Socher, R., Li, L.-J., Li, K., Fei-Fei, L.: ImageNet: a large-scale hierarchical image database. In: 2009 IEEE Conference on Computer Vision and Pattern Recognition, pp. 248–255 (2009). https://doi.org/10.1109/CVPR.2009.5206848
17. Loshchilov, I., Hutter, F.: Decoupled weight decay regularization (2019). arXiv: arXiv:1711.05101. https://doi.org/10.48550/arXiv.1711.05101
18. Kingma, D.P., Ba, J.: Adam: a method for stochastic optimization (2017). arXiv: arXiv:1412.6980. https://doi.org/10.48550/arXiv.1412.6980
19. Cervantes, J., Garcia-Lamont, F., Rodríguez-Mazahua, L., Lopez, A.: A comprehensive survey on support vector machine classification: applications, challenges and trends. Neurocomputing **408**, 189–215 (2020). https://doi.org/10.1016/j.neucom.2019.10.118
20. Cascante-Bonilla, P., Tan, F., Qi, Y., Ordonez, V.: Curriculum labeling: revisiting pseudo-labeling for semi-supervised learning. In: Proceedings of the AAAI Conference on Artificial Intelligence, vol. 35, no. 8, pp. 6912–6920 (2021). https://doi.org/10.1609/aaai.v35i8.16852
21. Rennie, J.D.M., Srebro, N.: Loss functions for preference levels: regression with discrete ordered labels (2005)
22. Pedregosa, F.: fabianp/mord. Python (2024). https://github.com/fabianp/mord. Accessed 27 Aug 2024

Evaluating Pretraining Strategies for OCT-Based Macular Degeneration Classification

Jessica Kächele[1,2,3(✉)], Robin Peretzke[1,2], Alexandra Ertl[1,2], Maximilian Fischer[1,2], Marlin Hanstein[1,4], Florian Max Hauptmann[1,5], Dimitrios Bounias[1,2], Marco Nolden[1,7], Peter Neher[1,3], and Klaus H. Maier-Hein[1,2,6,7]

[1] German Cancer Research Center (DKFZ) Heidelberg, Division of Medical Image Computing, Heidelberg, Germany
jessica.kaechele@dkfz-heidelberg.de
[2] Medical Faculty Heidelberg, University of Heidelberg, Heidelberg, Germany
[3] German Cancer Consortium (DKTK), DKFZ, Core Center Heidelberg, Heidelberg, Germany
[4] Faculty of Biology and Chemistry, Justus Liebig University, Giessen, Germany
[5] Ulm University of Applied Science (THU), Ulm, Germany
[6] Helmholtz Imaging, DKFZ, Heidelberg, Germany
[7] Pattern Analysis and Learning Group, Department of Radiation Oncology, Heidelberg, Germany

Abstract. Pretraining is a crucial step to improve the performance of deep learning models or to accelerate the training process. Ideal pretraining strategies enable fast adaptation to the target domain after pretraining and the development of foundation models. While in the natural scene image processing domain, recently various foundation models have been published, the medical domain is still lacking a general pretraining scheme, handling arbitrary acquisition modalities, diverse diseases, and varying anatomical structures. Current evaluations are mostly centered around common applications like organ segmentation in abdomen images or consider only a few selected pretraining strategies. In this paper, we compare various pretraining strategies, self-supervised and unsupervised on a classification task for longitudinal macular degeneration classification. We compare pretraining schemes specifically tailored towards the medical domain as well as schemes from the natural scene image domain. Our results show that upscaling pretraining schemes outweigh specific pretrained models in the medical or OCT scan domain. The code and hyperparameter settings can be found in our Github repository: https://github.com/MIC-DKFZ/mario.

Keywords: Pretraining Strategies · Macular Degeneration Classification · Siamese Networks · Self-Supervised Learning

---

J. Kächele, R. Peretzke, A. Ertl and M. Fischer—These authors contributed equally to this work.

1 Introduction

A common approach to enhancing deep learning model performance is pretraining, where the model is first trained on a large dataset before being fine-tuned for a specific task. Models pretrained on large-scale datasets consistently outperform those trained from scratch, even with a significant domain shift between the pretraining domain and the downstream task [8]. For example in the natural image domain, pretraining on large-scale datasets like ImageNet is widely applied and has even proven successful in more distantly related domains, such as medical pathology, where ImageNet pretrained models have demonstrated strong performance on specific medical tasks [3,6]. However, the medical domain is highly diverse, encompassing a variety of imaging modalities such as MRI, CT, or OCT, each with unique characteristics. It remains unclear whether pretraining on domain-specific medical data with limited scale can consistently outperform pretraining on large-scale datasets from unrelated domains when applied to specific medical tasks, such as predicting age-related macular degeneration (AMD) [4]. AMD is a progressive retinal disorder and a leading cause of visual impairment. Effective treatment relies heavily on accurately predicting disease progression using OCT scans [11].

In this paper, we evaluate a variety of pretraining strategies for predicting the progression of (AMD) in OCT scans, based on two distinct scenarios. First, we aim to classify the state of AMD progression (reduced, increased, stable, or uninterpretable) between two OCT scans of a patient taken at two different time points t_0 and t_1 (Task 1). Second, we seek to predict the further progression based solely on the OCT image from the time point t_0 (Task 2). Specifically, we compare the performance of models pre-trained on large-scale natural image datasets, such as ImageNet from the natural RGB domain, with foundation models pre-trained on medical datasets, such as RetFound and BiomedClip.

Additionally, we explore the effectiveness of pretraining directly on task-specific OCT data, using unsupervised strategies such as self-supervised contrastive learning (SSCP) and masked autoencoders (mAE). Our goal is to assess how different pretraining approaches impact the accuracy of disease progression predictions in AMD.

For Task 2, we aimed to artificially generate t_1 OCT scans using a CycleGAN architecture with ResNet50 weights for the generator encoder. The goal was to create synthetic t_1 OCT scans based on t_0 scans, and then generate further artificial t_0 scans from these synthetic t_1 scans. This approach was designed to capture disease-specific temporal information within the CycleGAN network while leveraging the ResNet50's ability to produce high-quality spatial features. However, this method did not achieve the desired success in solving the task, suggesting that more advanced techniques might be required to accurately capture temporal disease progression for predicting AMD.

2 Methods

This chapter outlines the methodological framework for predicting AMD progression, including a description of the network architectures for both downstream tasks, focusing on a Siamese network employed for Task 1 and also serving as the backbone for Task 2. Additionally, we detail the various pretraining strategies used for the network's encoder.

2.1 Model Architecture

For Task 1, as overall classification framework, we employed a Siamese network-based architecture [12], consisting of an encoder with shared weights for the OCT images X_{t0} and X_{t1}, a layer to combine the images' feature vectors by itemwise subtraction and two dense classification layers, mapping the combined feature vector to one of the four labels (reduced, increased, stable, or uninterpretable).

This design enables the close comparison of the extracted features and both time points. For Task 2, only X_{t1} was used as input to predict the disease progression. We leveraged the now fine-tuned encoder from Task 1 to extract AMD-specific features but trained a new classifier to accommodate the new task and class labels (reduce, stable, increase).

2.2 Pretraining Strategies for Task 1

We assessed various models and pretraining approaches for the encoder of our Siamese network, from general foundation models to dataset-specific pretraining strategies.

ImageNet. We initially utilized a Residual Network (ResNet) encoder pretrained on ImageNet, a dataset containing over 14 million natural images with associated classification labels for thousands of categories. Pretraining for classification on such a large and diverse dataset allows the model to learn a wide range of features, which can be leveraged for various tasks and have been successfully applied to medical downstream tasks in the past. Specifically, we use the ResNet50 implementation of the torchvision models module [9].

BiomedCLIP. BiomedCLIP [15] is a state-of-the-art vision-language foundation model tailored for the biomedical domain. It uses a Contrastive Language-Image pretraining (CLIP) [10] approach to learn a unified representation space from biomedical images and their associated textual annotations. BiomedCLIP was pre-trained on the PMC-15M dataset, which consists of 15 million image-text pairs from research articles on PubMed, covering a variety of 2D medical images including modalities such as X-ray, histopathology, or OCT scans. For our experiments, we used the vision encoder based on a vision transformer from Microsoft's public implementation on HuggingFace, *BiomedCLIP-PubMedBERT 256-vit base patch16 224*, and fine-tuned it for our specific downstream tasks.

RETFound. RETFound [16] is a foundation model develcped specifically for retinal imaging tasks. It is based on a masked autoencoder (MAE) pretraining strategy and uses a dataset of 1.6 million retinal images to learn a robust representation of relevant retinal features. They provide the weights of their Vision Transformer-based architecture for both color fundus images and OCT scans, the latter of which we used in our experiments.

Masked AutoEncoder Trained on OCT-Data. While ResNet50, pretrained on ImageNet can extract a wide range of meaningful features, it has not been specifically exposed to OCT data. To address this gap, we employed a masked autoencoder (MAE) to pre-train a model directly on OCT data, allowing it to learn relevant domain-specific features.

The mAE pretraining involves self-supervised training, where the input images are masked, and the model is tasked with reconstructing the missing parts. The mAE is trained on three different OCT datasets improving its generalization ability across different patient populations. The used datasets all together containing 89940 images are the MARIO challenge data, a Kaggle dataset [7], and OCT scans from the Optical Coherence Tomography Image Retinal Database [13] and include several pathologies such as diabetic-retinaopathy OCT-images or macular-hole OCT-scans.

The mAE architecture consists of a ResNet50 PretrainedResNetEncoder and a Decoder. The input images are masked with a ratio of 0.75. Skip connections at each stage preserve spatial details, which are later passed to the decoder. The decoder upsamples the encoded features using transposed convolutions and incorporates both skip connections and ResidualBlocks. The model is trained based on the MSE.

Self-supervised Contrastive Pretraining. Contrastive pretraining is a self-supervised method where the model learns to generate its own labels for images [1]. During pretraining, a batch of images includes an anchor image I^a and several negative samples I^-, which are all different from I^a. The model's encoder creates embeddings for each image, aiming to maximize the distance between the embeddings of I^a and each I^-. Cosine similarity is a common metric used to measure this distance. Besides negative samples, positive samples are needed for effective pretraining. These positive samples are generated by augmenting the anchor image I^a through transformations like color jittering, noise addition, cropping, or flipping. The embeddings of these augmented images I^{augm} should be close to the embedding of I^a. In self-supervised contrastive pretraining, only these augmented versions are used as positive samples, unlike supervised methods that use other images from the same class. We apply this SSCP approach in our experiments, which has shown promising results in medical applications [2] and for our experiments, we pretrain the ResNet50 architecture.

2.3 Cycle GAN Based OCT Scan Generation

Task 1 and Task 2 differ by the provided input data. While in Task 1 different time points of OCT scans are provided to predict AMD degeneration classification, Task 2 only provides initial OCT scans, without additional longitudinal time points. However, we aimed to re-use the Siamese architectures from Task 1 by incorporating a GAN model that generates synthetic OCT scancs of future time points. The initial time point and the artificial later time point are then processed similarly to Task 1 in the Siamese models. For training of the GAN model, we used a cycle GAN approach, since it seemed natural to assume direct connections between both images, since $X_{t0} = \Delta_t + X_{t1}$, where Δ_t represents the time between two OCT scans. For our experiments, we used the cycle GAN model from [17], which also offers the ResNet50 as the encoder part of the generator model. In our experiments from Task 1, we consistently observed superior results of the ResNet50 architecture compared to other models, thus we used ResNet50 as encoder part for the GAN training. We trained the model with pseudo-RGB images, similar to Task 1. During training, we set the first 8 latent channels to a fixed value Δ_t. For training of Task 2, we averaged all Δ_t from the whole dataset of Task 1 to conditionalize the cycle GAN to predict the OCT image of a later time point. An averaged Δ_t approximates the passing time between a OCT follow up scan.

3 Results

This section describes the convergence of the OCT-based pretraining with the mAE and the SSCP and the results for Tasks 1 and 2 based on different pretraining strategies.

3.1 Masked Autoencoder

The mAE converged after around 60 epochs and sufficiently minimized the MSE on both the training and validation set. An example of the masked input OCT image with prediction and mask is shown in Fig. 1. The encoder, fine-tuned on OCT images is then used as the encoder backbone for the Siamese network.

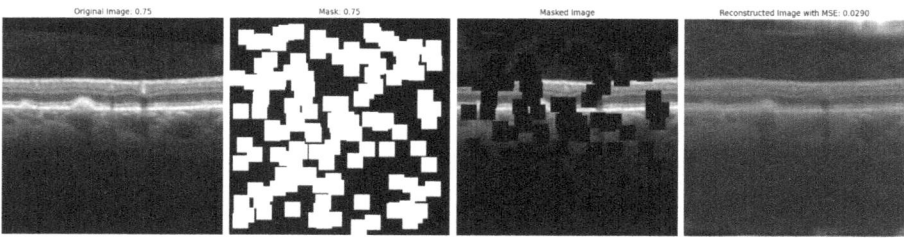

Fig. 1. Randomly selected OCT images, which is masked and fed to the mAE which targets to reconstruct the masked parts

3.2 Self-supervised Contrastive Pretraining

During contrastive pretraining, we used random noise, vertical and horizontal flips, cropping, and intensity based augmentations. We trained the model for 100 epochs and after 30 epochs no further substantial reduction of the cosine similarity that we used as loss function was achieved.

3.3 Generating Synthetic OCT Scans

Our experiments for generating synthetic OCT images show that temporal components in OCT scans can be partially predicted. However, for very different images X_{t0} and X_{t1}, our results show only little similarity between the artificial X_{t1} and the ground truth (Fig. 2).

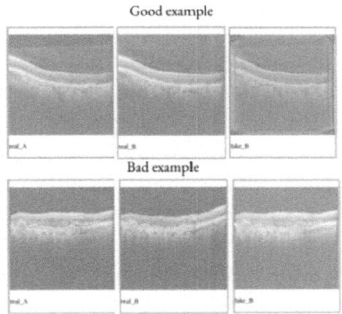

Fig. 2. Example of synthetic OCT images. Information about Δ_t was encoded in the first channels, which results in a red image impression.

3.4 Classification Results

Our findings demonstrate that pretraining enhances the model performance (see Table 1). While we expected superior performance of domain specific pretrained models, like RETFound, this OCT tailored model did not outperform Biomed-Clip. Also the other domain targeted aproaches (mAE and contrastive) could not compete with BiomedClip. Our results show that the performance on Task 2 was significantly inferior to that on Task 1, as evidenced by Table 1.

4 Discussion

In our experiments for Task 1, we found that pretraining can indeed be beneficial. This finding is consistent with the prevailing literature, which asserts the efficacy of pretrained models in enhancing performance on domain-specific tasks.

Table 1. Cross-validation results on the train set. The best models on the trainset were then also trained on the combined train and validation data, which is denoted in brackets.

Model	Task 1 [F1]		Task2 [F1]	
	Macro	Micro	Macro	Micro
ResNet50 (No pretraining)	0.619	0.755		
ResNet50 (ImageNet weights)	0.627 (0.654)	0.756 (0.775)	0.197	0.445
BiomedClip (ViT)	**0.634** (0.649)	**0.767** (0.774)	0.311	0.752
RETFound (ViT)	0.629 (0.633)	0.760 (0.760)	0.369	0.609
mAE	0.624	0.757		
Contrastive pretraining	0.578	0.726		
BiomedCLIP with GAN			0.284	0.744

However, contrary to our initial hypothesis, our results suggest that domain-specific pretraining does not confer additional advantages over general pretraining methods. Both approaches delivered comparable performance metrics, suggesting that the increased effort required to collect application-specific datasets may not be justifiable and that additional scalability can accommodate domain shifts. An important consideration for future research is finding the optimal balance between domain proximity and scalability, especially until a comprehensive large-scale radiology foundation model becomes available. Such a model could significantly reduce the need to develop task-specific deep learning models in the medical domain.

The findings from Task 2 delve into the complexities of predicting disease progression from a single timepoint. For this task, we aimed to use a Cycle-GAN to predict future OCT scans (t_1) based on the t_0 scans, with the goal of learning disease-specific temporal information. We employed the top-performing pre-trained method from Task 1, ResNet50, as the encoder for the generator to maintain high-quality spatial feature generation. Although the CycleGAN generated synthetic future OCT scans, feeding these into our Siamese network did not yield better results compared to directly inputting the t_0 OCT scan into a classification network. This indicates that the CycleGAN were not able to successfully learn meaningful temporal disease-specific information based on the t_0 OCT scan.

Our experiments revealed that conventional classification models are inadequate for this purpose due to their limitations in handling temporal data. To address this shortfall, we propose the integration of more advanced methodologies such as utilizing ODESolve, as demonstrated in [14], or adopting a recalibration model [5].

References

1. Chen, T., Kornblith, S., Norouzi, M., Hinton, G.: A simple framework for contrastive learning of visual representations. In: International Conference on Machine Learning, pp. 1597–1607. PMLR (2020)
2. Ciga, O., Xu, T., Martel, A.L.: Self supervised contrastive learning for digital histopathology. Mach. Learn. Appl. **7**, 100198 (2022). https://doi.org/10.1016/j.mlwa.2021.100198. https://www.sciencedirect.com/science/article/pii/S2666827021000992
3. Frid-Adar, M., Ben-Cohen, A., Amer, R., Greenspan, H.: Improving the segmentation of anatomical structures in chest radiographs using U-Net with an imagenet pre-trained encoder. In: Image Analysis for Moving Organ, Breast, and Thoracic Images: Third International Workshop, RAMBO 2018, Fourth International Workshop, BIA 2018, and First International Workshop, TIA 2018, Held in Conjunction with MICCAI 2018, Granada, Spain, 16 and 20 September 2018, Proceedings 3, pp. 159–168. Springer (2018)
4. Kataria, T., Knudsen, B., Elhabian, S.: To pretrain or not to pretrain? A case study of domain-specific pretraining for semantic segmentation in histopathology. In: Medical Image Learning with Limited and Noisy Data, pp. 246–256. Springer (2023)
5. Konwer, A., Xu, X., Bae, J., Chen, C., Prasanna, P.: Temporal context matters: enhancing single image prediction with disease progression representations. In: 2022 IEEE/CVF Conference on Computer Vision and Pattern Recognition (CVPR), Los Alamitos, CA, USA, pp. 18802–18813. IEEE Computer Society (2022)
6. Li, X., Cen, M., Xu, J., Zhang, H., Xu, X.S.: Improving feature extraction from histopathological images through a fine-tuning ImageNet model. J. Pathol. Inform. **13**, 100115 (2022)
7. Mooney, P.: Retinal OCT images (optical coherence tomography). Kaggle dataset (2018)
8. Morid, M.A., Borjali, A., Del Fiol, G.: A scoping review of transfer learning research on medical image analysis using ImageNet. Comput. Biol. Med. **128**, 104115 (2021)
9. Paszke, A., et al.: Torchvision: models and transforms for image and video (2019). https://pytorch.org/vision/stable/models.html. Accessed 30 Aug 2024
10. Radford, A., et al.: Learning transferable visual models from natural language supervision (2021). https://arxiv.org/abs/2103.00020
11. Rasmussen, A., Sander, B.: Long-term longitudinal study of patients treated with ranibizumab for neovascular age-related macular degeneration. Curr. Opin. Ophthalmol. **25**(3), 158–163 (2014)
12. Rivail, A., et al.: Modeling disease progression in retinal OCTs with longitudinal self-supervised learning. In: Predictive Intelligence in Medicine, pp. 44–52. Springer (2019)
13. Subramanian, M., Shanmugavadivel, K., Naren, O.S., Premkumar, K., Rankish, K.: Classification of retinal OCT images using deep learning. In: 2022 International Conference on Computer Communication and Informatics (ICCCI), pp. 1–7. IEEE (2022)
14. Zeghlache, R., et al.: LMT: longitudinal mixing training, a framework to predict disease progression from a single image. In: Machine Learning in Medical Imaging: 14th International Workshop. MLMI 2023, Held in Conjunction with MICCAI 2023, Vancouver, BC, Canada, 8 October 2023, Proceedings, Part II, pp. 22–32. Springer, Heidelberg (2023)

15. Zhang, S., et al.: BiomedCLIP: a multimodal biomedical foundation model pretrained from fifteen million scientific image-text pairs (2024). https://arxiv.org/abs/2303.00915
16. Zhou, Y., et al.: A foundation model for generalizable disease detection from retinal images. Nature **622**(7981), 156–163 (2023). https://doi.org/10.1038/s41586-023-06555-x. https://www.nature.com/articles/s41586-023-06555-x
17. Zhu, J.Y., Park, T., Isola, P., Efros, A.A.: Unpaired image-to-image translation using cycle-consistent adversarial networks. In: 2017 IEEE International Conference on Computer Vision (ICCV) (2017)

Classification and Prediction of Age-Related Macular Degeneration Progression Using OCT Images and Multiple Instance Learning

Alberto J. Beltrán-Carrero[1](✉)[iD], Javier Torresano-Rodríguez[2], Esther Santos-Vicente[2], María J. Aparicio Hernández-Lastras[2], Álvaro Caballero-Sastre[1], María J. Ledesma-Carbayo[1,3][iD], and Juan J. Gómez-Valverde[1,3][iD]

[1] Biomedical Image Technologies (BIT), ETSI Telecomunicación, Universidad Politécnica de Madrid, Madrid, Spain
aj.beltran.carrero@upm.es
[2] Ophthalmology Service of the Provincial Ophthalmic Institute, Hospital Universitario Gregorio Marañón, Madrid, Spain
[3] Centro de Investigación Biomédica en Red de Bioingeniería, Biomateriales y Nanomedicina (CIBER-BBN), Madrid, Spain

Abstract. The study introduces a novel approach for classifying and predicting the progression of Age-related Macular Degeneration (AMD) using Optical Coherence Tomography (OCT) images and Multiple Instance Learning (MIL). AMD is a leading cause of vision impairment worldwide, making effective monitoring and treatment essential, particularly with anti-VEGF therapy. However, the increasing number of patients and the frequency of follow-up visits pose challenges for healthcare systems. This approach addresses two key tasks: (1) classifying changes between consecutive 2D OCT B-scans and (2) predicting disease progression within a 3-month period. For task 1, the model incorporates contextual information from adjacent B-scans and applies bidirectional cross-attention to learn time-dependent features. For task 2, a MIL-based architecture is used to identify the most significant slices within an OCT volume. The results demonstrated the effectiveness of the proposed methods. In task 1, the model achieved a mean score of 0.7488 across all evaluation metrics. For task 2, the mean score was 0.4478, reflecting the complexity of disease progression prediction. This approach offers improvements over baseline models and contributes to the development automated tools for AMD management, potentially easing the burden on ophthalmology services and improving personalized patient care.

Keywords: AMD · OCT · Anti-VEFG · MIL

1 Introduction

Age-related macular degeneration (AMD) is a chronic, progressive retinal disease that affects the macula, the area of the retina responsible for the sharpest vision. As the leading cause of irreversible vision loss in the elderly population in developed countries, AMD is a significant public health concern, currently impacting approximately 196 million people worldwide [1]. In recent years, the introduction of anti-vascular endothelial growth factor (anti-VEGF) therapy for neovascular AMD has marked a major advancement in treatment, significantly improving outcomes [2]. However, the growing number of patients requiring treatment and frequent follow-ups has placed considerable strain on healthcare resources, resulting in substantial costs and the risk of overburdening ophthalmology services [3].

In this context, developing comprehensive solutions to optimize patient management workflows is crucial. By integrating clinical and imaging biomarkers, the diagnosis, treatment, and follow-up of AMD patients could become significantly more efficient [4,5]. A paradigm shift could be achieved through the application of deep learning algorithms, which can predict the need for anti-VEGF treatment and assist in monitoring disease progression as well as evaluating therapeutic efficacy [6]. The implementation of these technologies would not only reduce the burden on healthcare systems by enhancing resource allocation and reducing follow-up intervals, but also provide more personalized and timely interventions for patients [7].

Several studies have explored the application of machine learning and deep learning techniques to assist in the monitoring of AMD, most of these efforts focus on predicting the progression from early or intermediate non-exudative stages to advanced exudative stages [8,9], as well as forecasting treatment requirements [10–12]. Despite previous efforts, there remains an unmet need for accurate prediction of AMD progression in patients closely monitored on anti-VEGF treatment plans. The Monitoring Age-related Macular Degeneration Progression in Optical Coherence Tomography (MARIO) challenge at MICCAI 2024 aims to address this gap by evaluating both existing and novel algorithms for detecting progression of neovascular activity in OCT scans of patients with exudative AMD, with the ultimate goal of improving treatment planning.

The challenge is divided into two tasks. Task 1 focuses on pairs of 2D B-scans from two consecutive OCT scans, with the goal of classifying the changes between these slices, which are typically compared side by side by clinicians. Task 2 shifts the focus to predicting future disease progression within a 3-month period for patients undergoing anti-VEGF treatment based on 2D layers.

This work presents a novel approach that addresses both tasks, using two architectures inspired by Multiple Instance Learning (MIL) to effectively exploit the contextual information of successive B-scans.

2 Methods

2.1 Task 1: Classify Evolution Between Two Pairs of 2-D Slices from Two Consecutive 2D OCT Acquisitions

Dataset. Data for task 1 consisted in cases of 2 OCT B-scans and 2 fundus images associated with 2 consecutive visits. Along with image information, patient identifier, side eye, sex, age, visit number and the days between the two visits are also provided. Labels are given at B-scan level, that is, B-scans with the same fundus image associated (from the same acquisition) can have different labels. In this task, the following types of change in activity were defined: reduced (0), stable (1), worsened (2) and other (3).

Pre-processing. Original data samples consisted of a pair of individual OCT B-scans, the associated fundus, the clinical variables and the label assigned to that pair of B-scans. For the purposes of our method, we have only considered the B-scan pairs from t_i and t_{i+1}. First of all, we created the entire OCT volumes of each exam, by selecting all the B-scans that have associated the same fundus identifier, which is also associated with a single volume acquisition. Then, each sample consisted in a pair of OCT volumes and the labels associated to their B-scans pairs. To achieve this, we ensured that during volume creation all B-scans were correctly ordered, using the B-scan indexes provided in the original dataset. Therefore, we took the provided labels in the same order and created a list of labels, which were used as ground truth for each volume during training. Finally, we computed the mean and standard deviation of the volumes and performed intensity normalization using these values.

Method. We propose a model which aims to leverage the contextual information from adjacent B-scans within an OCT volume and over time. The overall architecture is illustrated in Fig. 1. We hypothesise that having contextual information might be potentially useful for the model, in order to better understand the activity of individual B-scans based on their neighbourhood. First, a batch of N 2-D slices extracted from an OCT volume was processed by a vision transformer image encoder. The initial weights of this encoder were taken from the RETFound OCT foundation model [13], which has shown remarkable performance in several retinal-related disease classification tasks and whose weights are publicly available. During training, we only freeze the patch-embedding layer of the encoder due to GPU memory limitations.

From each B-scan, we extracted the corresponding CLS token from the encoder and we stacked all the vectors from each time step, separately. Both vector matrices were then fed into a bidirectional cross-attention module, which produced a pair of attention matrices. This module was designed to process the information from all B-scans from both time steps together. In this way, we forced the model to learn the dependencies between two groups of consecutive slices over time. The module followed a similar mechanism to the cross-attention module of the transformer decoder [14]. It had three weight matrices for each

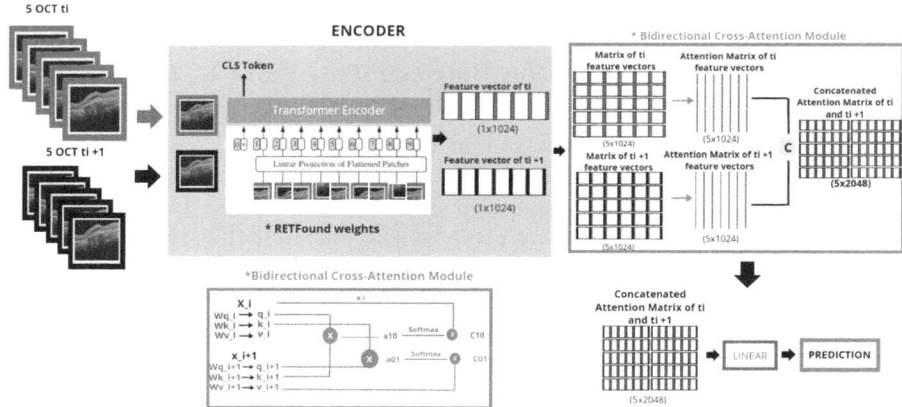

Fig. 1. Proposed architecture for classifying the type of change in activity between two consecutive OCT exams (task 1). The model is designed to leverage the context information from a window of 5 consecutive B-scans and the dependencies between B-scans from different time steps. Red arrows refer to the bidirectional cross-attention operation. (Color figure online)

time step: W_{qi}, W_{ki}, W_{vi}. W_{qi} referred to the matrix that produced the query for time step i, while W_{ki} and W_{vi} produced the keys and values, respectively. Then, two context matrices were computed according to Eq. 1. Since these matrices contain cross-time dependencies for each B-scan, they were concatenated along the feature axis and, finally, the resulting matrix was passed through a linear layer, which performed the classification for each individual B-scan. It is worth noting that the linear layer generated a prediction for each row of the matrix, i.e. for each B-scan, since we cannot assume that all slices from a close neighbourhood would have the same label.

$$C_{i+1,i} = Softmax\left(\frac{Q_i K_{i+1}^T}{\sqrt{d_k}}\right) V_i, \quad C_{i,i+1} = Softmax\left(\frac{Q_{i+1} K_i^T}{\sqrt{d_k}}\right) V_{i+1} \quad (1)$$

Given that the model is trained with random batches of N consecutive B-scans and that not all volumes have the same length, during inference time we had to implement a method to generate the predictions in batches of N B-scans. We simply divided the volumes into groups of N slices and, for those cases where the volume length was not divisible by N, the remaining slices were processed together at the end. This means that for some volumes, the predictions for the last B-scans were computed using a context window smaller than N.

2.2 Task 2: Prediction of Evolution Within 3 Months of AMD on OCT 2D Slices for Planning Treatment Anti-VEGF

Dataset. Data for task 2 consisted in cases of individual OCT B-scans and the corresponding fundus image, both associated with a single visit. Along with

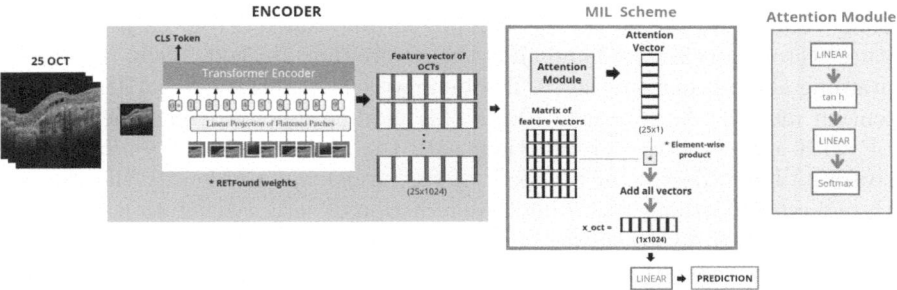

Fig. 2. Proposed architecture for predicting the type of change in activity of an OCT exam (task 2). The model is able to leverage the context information of N consecutive B-scans by following a Multiple Instance Learning (MIL) scheme.

image information, patient identifier, side eye, sex, age and visit number were also provided. For this task, while labels were given at the B-scan level, B-scans with the same fundus image associated (from the same exam) had the same label, i.e. we assumed that we had exam-level labels. In this task, the following types of change activity were defined: reduced (0), stable (1) and increased (2).

Pre-processing. The original samples consisted of a single B-scan, the associated fundus, the clinical variables and the assigned label. For the purposes of our method, we have only considered the OCT B-scans. We needed to create OCT volumes of the same size, by taking all B-scans with the same associated fundus identifier (unique for each volume acquisition). However, we noticed that the number of B-scans per volume varied across the dataset. While most of them were similar in size (19 and 25 B-scans/volume), there were volumes with smaller and larger size. To solve this issue without discarding samples, we decided to create the volumes with a constant size of 25 B-scans/volume. First, we selected the middle B-scan of the entire acquisition and we took the previous and consecutive 12 B-scans. For volumes with less than 25 B-scans, we repeated the extreme slices to complete the volume. In this way, we were able to create a dataset with volumes of the same size. Regarding the labels, we noticed that all the B-scans from the same volume were assigned the same label, thus, we can use it as a label for the whole volume. Finally, as in task 1, we computed the mean and standard deviation of all volumes, in order to perform intensity normalization.

Method. We propose a model architecture that integrates the information from N OCT B-scans from the same volume to produce a global label for all B-scans included. The architecture is illustrated in Fig. 2. First of all, a batch of N consecutive B-scans extracted from a volume was fed to the image encoder. As in task 1, the encoder consisted in a vision transformer whose initial weights were taken from the RETFound OCT model [13] and, this time, no layers were freeze during training, so the encoder is completely fine-tuned. From each B-scan, a feature vector was extracted and all of them were then stacked. Then, the matrix containing all vectors from the OCT volume were passed through a MIL-oriented attention module. As shown in Fig. 2, this module followed the structure of a MLP with a *tanh* activation function, with the addition of a *Softmax* as final activation. The objective of this part of the model was to identify the B-scans with highest importance regarding the prediction of activity evolution. Since the B-scans of an entire volume were equally labelled, that is, the type of change in activity is associated to the whole exam, we assumed that there were regions of the volume more relevant to the identification of change activity, which aligned better with the nature of the disease activity, given that typical biomarkers, such as retinal fluids, are not uniformly distributed across all the volume. Therefore, the attention module learned to produce a vector that assigns an importance value to each slice. Once this vector is computed, the matrix of feature vectors is multiplied by the attention scores. Subsequently, the matrix is reduced to a single vector by adding of rows, which means adding values on the feature axis. This vector is then fed into the linear layer that produces the prediction. It should be noted that, in this case, the model generated an individual label for the entire OCT volume. Since not all volumes have the same number of slices, this label was extended to the corresponding original number of B-scans of the input volume.

2.3 Training

Training Data. After conducting the aforementioned pre-processing steps, a total of 573 OCT volumes for task 1 and 330 for task 2 comprised the final training dataset. All volumes underwent intensity scaling and normalization before training. Also, they were resized to $[N, 224, 224]$ (N being the number of slices), in order to match the 2-D input size in which the image encoder was pre-trained. Then, we implemented a set of data augmentation operations using MONAI framework [15], which were conducted during model training. These operations included: random flip in all dimensions, random intensity scaling and random Gaussian noise addition. Regarding task 1, we ensured that the same augmentation operations were applied to paired volumes.

Moreover, for task 2, we implemented weighted random sampling during training. Weighted random sampling is a method which aims to balance the contribution of all classes to model training, by assigning the probability of taking a sample from one class based on the proportion of that class in the dataset, resulting in samples from majority classes being taken less often than those from minority classes. In our case, for each sample belonging to class c, its weight value corresponds to $w_c = N/N_c$, where N refers to the total number of samples in the dataset, and N_c refers to the number of samples in the dataset belonging to class c. This method helps the model to achieve better generalization to all classes without discarding samples. Nonetheless, this method can only be employed for task 2, since we have a single label per sample. In task 1, as commented before, a

Table 1. Evaluation metrics for task 1.

Model	F1-Score	Rk-corr. coef.	Specificity	Mean
RETFound [13]	0.7476	0.5113	0.8857	0.7149
Ours ($N = 5$)	**0.7849**	**0.5680**	**0.8936**	**0.7488**

sample of N B-scans taken from a volume can potentially have more than one label, so sample weighting becomes useless.

Optimization and Hardware. All experiments were conducted using the same optimization scheme. We employed the AdamW optimizer with a weight decay of 0.05 and $\beta_1 = 0.9, \beta_2 = 0.95$. We used a cosine learning rate scheduler with an initial learning rate of 0.0001 and 10 warm-up epochs. The total number of epochs was set to 100. We trained all our models in NVIDIA A100 40 GB GPUs provided by Magerit-3 cluster from the Centro de Supercomputación y Visualización de Madrid (CesViMa).

In order to select the best performing model from each experiment, we compute the mean of all evaluation metrics (presented in the following section) at the validation phase, which is conducted after every training epoch. Then, we saved the model with the global best mean metric.

2.4 Evaluation

The evaluation data comprised 7010 cases for task 1 and 3822 cases for task 2. As described before, for task 1 our model performed n-slice inference in order to generate a label for each B-scans pair. For task 2, we performed inference at volume level and the resulting label is assigned to all volume B-scans.

The performance metrics employed for model evaluation included: F1-score, Rk-correlation coefficient, Quadratic-weighted Kappa (for task 2 only) and Specificity. The implementation of these metrics was provided by the organisers.

3 Results

3.1 Task 1

Evaluation metrics and the confusion matrices for the experiments of task 1 are presented in Table 1 and Fig. 3. Our proposed model is compared with the performance of the RETFound model as a reference, which was also our best submission during the development phase. In this approach, we directly trained the vision transformer backbone to predict the label for a pair of individual B-scans.

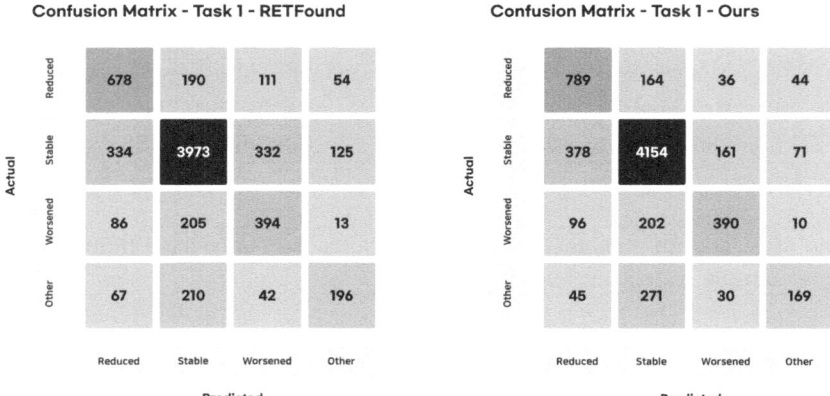

Fig. 3. Confusion matrices of best models for task 1. (Left) RETFound. (Right) Our model.

3.2 Task 2

Evaluation metrics and the confusion matrices for the experiments of task 2 are presented in Table 2 and Fig. 4. We have compared our model with respect to the RETFound and a ResNet 3D architecture. The RETFound was trained to predict the label of individual B-scans. The ResNet architecture is trained to predict the label of entire volumes. We also included the results of our proposed model trained using fixed-size volumes of 25 B-scans and using the complete original volumes.

Table 2. Evaluation metrics for task 2.

Model	F1-Score	Rk-corr. coef.	QW-Kappa	Specificity	Mean
RETFound [13]	0.7015	0.0262	0.0078	0.6721	0.3519
ResNet50-3D	0.7002	0.1330	**0.2422**	0.6925	0.4420
Ours (25 slices)	0.7310	**0.1736**	0.1874	**0.6990**	**0.4478**
Ours (all slices)	**0.7412**	0.1571	0.1520	0.6868	0.4343

4 Discussion

The main contribution of our methods was based on the assumption that the contextual information provided by adjacent B-scans would be useful in training a model for predicting the type of change in activity at the B-scan level. The results we obtained confirmed this assumption for both tasks showing that a model can effectively benefit from processing groups of consecutive B-scans and

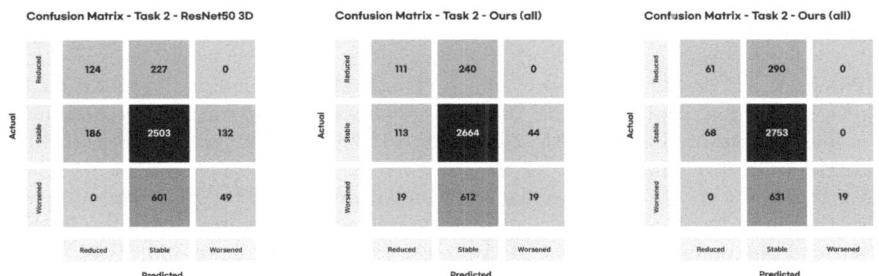

Fig. 4. Confusion matrices of best models for task 2. (Left) ResNet50 3D. (Center) Our model using volumes of 25 slices. (Right) Our model using all slices from original volumes.

improve its performance, compared to using only individual slices. In particular, we noticed such improvement for task 2, where models that leverages context information, by either using neighbouring slices or 3-D convolutions, clearly outperform a model which only uses individual B-scans for training. With respect to task 1, our bidirectional cross-attention module has shown that it is able to integrate information across B-Scans and time instants. Compared to other simpler approaches, for instance, recurrent neural networks, our module offers scalability and computing efficiency, as it is based on the Transformer's attention mechanism.

Our approach does not integrated other information (fundus and clinical variables) which can potentially help to improve the performance. Additionaly, we did not include any synthetic data augmentation method to handle the heavy class imbalance in the datasets, which is likely to provide better model generalization. Future lines of investigation should focus on integrating information from other sources and developing data effective augmentation techniques.

Code Repository: The code for this project is available on GitHub: https://github.com/BIT-UPM/mario_miccai_24_step_amd.

Acknowledgments. The authors acknowledge the support of Ministerio de Ciencia e Innovación, Agencia Estatal de Investigación, under grants TED2021-131951B-I00 and PID2022-141493OB-I00 (10.13039/501100011033/MCIN/AEI/ERDF, UE), cofinanced by European Regional Development Fund (ERDF), 'A way of making Europe' and the Next Generation EU funds. AJBC is supported by a FPI grant from the Ministerio de Ciencia e Innovación - PREP2022-000162. The authors gratefully acknowledge the Universidad Politécnica de Madrid (www.upm.es) for providing computing resources on Magerit Supercomputer.

Disclosure of Interests. The authors have no competing interests to declare that are relevant to the content of this article.

References

1. Wong, W.L., et al.: Global prevalence of age-related macular degeneration and disease burden projection for 2020 and 2040: a systematic review and meta-analysis. Lancet Glob. Health **2**(2), e106–e116 (2014). https://doi.org/10.1016/S2214-109X(13)70145-1
2. Finger, R.P., et al.: Anti-vascular endothelial growth factor in neovascular age-related macular degeneration-a systematic review of the impact of anti-VEGF on patient outcomes and healthcare systems. BMC Ophthalmol. **20**, 1–14 (2020). https://doi.org/10.1186/s12886-020-01554-2
3. Ruiz-Moreno, J.M., Arias, L., Abraldes, M.J., Montero, J., Udaondo, P., The RAMDEBURS Study Group: Economic burden of age-related macular degeneration in routine clinical practice: the RAMDEBURS study. Int. Ophthalmol. **41**(10), 3427–3436 (2021). https://doi.org/10.1007/s10792-021-01906-x
4. Guymer, R., Wu, Z.: Age-related macular degeneration (AMD): more than meets the eye. The role of multimodal imaging in today's management of AMD-a review. Clin. Exp. Ophthalmol. **48**(7), 983–995 (2020). https://doi.org/10.1111/ceo.13837
5. Schmidt-Erfurth, U., Waldstein, S.M.: A paradigm shift in imaging biomarkers in neovascular age-related macular degeneration. Prog. Retin. Eye Res. **50**, 1–24 (2016). https://doi.org/10.1016/j.preteyeres.2015.07.007
6. Schmidt-Erfurth, U., et al.: Ai-based monitoring of retinal fluid in disease activity and under therapy. Prog. Retin. Eye Res. **86**, 100972 (2022). https://doi.org/10.1016/j.preteyeres.2021.100972
7. Tan, T.E., Wong, T.Y., Ting, D.: Artificial intelligence for prediction of anti-VEGF treatment burden in retinal diseases: towards precision medicine. Ophthalmol. Retina **5**(7), 601–603 (2021). https://doi.org/10.1016/j.oret.2021.05.001
8. Banerjee, I., et al.: A deep-learning approach for prognosis of age-related macular degeneration disease using SD-OCT imaging biomarkers. arXiv preprint arXiv:1902.10700 (2019). https://doi.org/10.48550/arXiv.1902.10700
9. Liu, Y., et al.: Prediction of OCT images of short-term response to anti-VEGF treatment for neovascular age-related macular degeneration using generative adversarial network. Br. J. Ophthalmol. **104**(12), 1735–1740 (2020). https://doi.org/10.1136/bjophthalmol-2019-315338
10. Bogunović, H., et al.: Prediction of anti-VEGF treatment requirements in neovascular AMD using a machine learning approach. Invest. Ophthalmol. Vis. Sci. **58**(7), 3240–3248 (2017). https://doi.org/10.1167/iovs.16-21053
11. Gallardo, M., et al.: Machine learning can predict anti-VEGF treatment demand in a treat-and-extend regimen for patients with neovascular AMD, DME, and RVO associated macular edema. Ophthalmol. Retina **5**(7), 604–624 (2021). https://doi.org/10.1016/j.oret.2021.05.002
12. Romo-Bucheli, D., Erfurth, U.S., Bogunović, H.: End-to-end deep learning model for predicting treatment requirements in neovascular AMD from longitudinal retinal oct imaging. IEEE J. Biomed. Health Inform. **24**(12), 3456–3465 (2020). https://doi.org/10.1109/JBHI.2020.3000136
13. Zhou, Y., Chia, M.A., et al.: A foundation model for generalizable disease detection from retinal images. Nature **622**(7981), 156–163 (2023). https://doi.org/10.1038/s41586-023-06555-x
14. Vaswani, A., et al.: Attention is all you need (2023). https://arxiv.org/abs/1706.03762
15. Jorge Cardoso, M., Li, W., et al.: Monai: an open-source framework for deep learning in healthcare (2022). https://arxiv.org/abs/2211.02701

Multi-modal Siamese Ensemble for Neovascular AMD Classification and Prediction from Optical Coherence Tomography

Sebastien Richard(✉) and Marie Beurton-Aimar

LaBRI UMR 5800, Bordeaux University, Bordeaux, France
sebastien.richard@labri.fr

Abstract. Age-related Macular Degeneration (AMD) is a leading cause of legal blindness worldwide, with late-stage forms requiring close monitoring for effective treatment. Currently, this monitoring is performed manually by clinicians using Optical Coherence Tomography (OCT), a process that is both labor-intensive and time-consuming. Computer-assisted diagnosis models provide a scalable alternative. In this work, we propose a method for the MARIO challenge based on an Ensemble of Siamese networks to analyze two consecutive OCT B-scans and classify AMD progression. Our approach combines three state-of-the-art CNNs and incorporates categorical patient data and a 2D infrared OCT image within a customized fusion module. Additionally, we use Knowledge Distillation to train a neural network to predict disease evolution from a single B-scan, addressing the need for predicting future progression from one time point. Experimental results show the validity of our method for the classification task, achieving an F1-score of 0.832 on an external evaluation set.

Keywords: Age-related Macular Degeneration · Ensemble model · Multimodal · Siamese

1 Introduction

Context. Age-related Macular Degeneration (AMD) is a multifactorial eye disease characterized by the progressive degradation of the macular region of the retina [20]. In Europe, about 12% of people aged 45 and older are affected by AMD, with the number of cases worldwide expected to reach 280 million by 2040 [31]. Late-stage AMD is categorized into two distinct forms: atrophic and neovascular. The neovascular form is especially aggressive, characterized by the rapid proliferation of abnormal blood vessels beneath the macula, which can result in hemorrhages and fluid leakage [8].

Current treatment for neovascular AMD primarily relies on anti-VEGF therapies, which target a cytokine that drives angiogenesis and increases vascular permeability. While effective, these therapies require frequent intravitreal injections and close monitoring, placing a significant burden on both patients and healthcare systems [4,22]. Developing cost-efficient Computer-Assisted Diagnosis (CAD) methods could improve patient care by making examinations scalable and reproducible.

The MARIO challenge is structured into two distinct but related tasks. The objective of the first task is to classify the evolution of neovascular AMD between a pair of 2-D slices from two consecutive Optical Coherence Tomography (OCT) acquisitions. The objective of the second task is to predict the evolution of neovascular AMD within a three-month period following a given examination, based on a single 2-D slice from one OCT acquisition.

In this paper, we propose our solution for both tasks using a multi-modal Siamese Ensemble model. Our approach integrates three state-of-the-art Convolutional Neural Networks (CNNs) as submodels. We report our results on our own test set as well as on the external evaluation set that was provided for ranking the participants during the development phase of the challenge.

2 Method

Convolutional Neural Networks (CNNs) are a type of neural network specifically designed for computer vision tasks. Inspired by the human visual system, CNNs use convolutional layers to extract features from input data [7]. Unlike traditional neural networks, CNNs process 2D images directly and exhibit remarkable properties such as local connectivity and translation invariance [19]. In medical imaging, CNNs outperform older methods based on manually designed features and expert knowledge, successful in tasks such as image segmentation, computer-aided diagnosis, disease detection and classification, and medical image retrieval [3].

2.1 Ensemble Model

Ensemble models in Deep Learning (DL) improve predictive performance by combining outputs from multiple independent neural networks. It usually results in lower squared error than any single model alone [9]. This is particularly useful for small datasets or complex problems where optimizing a single model is challenging. Each model in an Ensemble starts from a different point, increasing the likelihood of finding near-optimal parameters.

For this challenge, we implemented these submodels as Siamese networks. Introduced by Bromley et al. (1994), Siamese networks are designed for tasks requiring input comparison [6]. The architecture consists of two identical, weight-sharing networks that process different data instances in parallel. The resulting output features are often compared, for example using distance metrics. Siamese networks are widely used in medical imaging, including radiographs, MRI, and fundus images [1,14,15,23].

Our approach involves three main steps. First, three Siamese CNN models are trained to process pairs of B-scans as input (Fig. 1) (1)), with each Siamese linked to a module incorporating categorical patient data. Next, an Ensemble module is trained to combine the outputs of the three Siamese networks, using another CNN that processes the 2D infrared localizer image to improve ensembling (Fig. 1) (2)). Finally, for Task 2, three new CNN submodels are trained using Knowledge Distillation, with the pretrained Siamese networks serving as the teacher. The CNNs were chosen for their popularity and recent adoption. Below is a brief overview of each.

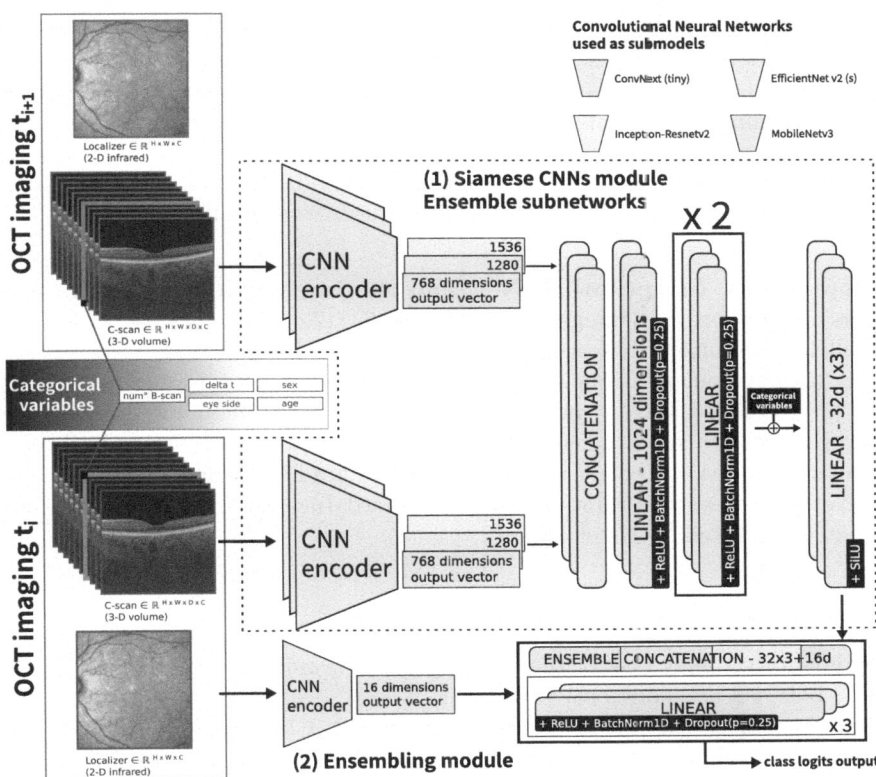

Fig. 1. This figure shows the architecture of our **Multi-modal Siamese Ensemble**. In the first phase, three Siamese networks process pairs of B-scans. Their concatenated feature vectors are passed through fully-connected layers, where categorical data about the patients are added. In a second phase, an Ensemble of these networks is trained, with a MobileNetv3 backbone to process the localizer image. Finally, the resulting feature vectors are concatenated and passed through a clssification head to generate the final classification output.

ConvNeXt: A ConvNet for the 2020s (Liu et al. 2022) [17]. ConvNeXt is inspired by the success of Vision Transformers (ViTs), which have outperformed CNNs in various tasks. Interestingly, the most successful ViT variants, such as Swin Transformer, have reintroduced some CNN principles. The authors aim to identify the properties that are responsible for Swin Transformer performance, integrate these features into a purely convolutional architecture, and try to achieve comparable results. Starting with a ResNet model, they progressively adapt it towards a ViT-like design.

EfficientNet: Rethinking Model Scaling for Convolutional Neural Networks (Tan and Quoc 2020) [26]. The EfficientNet model was proposed as a solution to automatically select the optimal number and size of layers of a model, achieving a balance between performance and computational cost. The authors look to optimize in parallel the depth, width, and resolution, and shows that better accuracy is achieved by scaling each of these parameters by a constant ratio, referred to as the compound scaling method. Its main building block is mobile inverted bottleneck MBConv with squeeze-and-excitation.

Inception-v4, Inception-ResNet, and the Impact of Residual Connections on Learning (Szegedy et al. 2016) [25]. The Inception block uses parallel convolutions to capture diverse features from various receptive field sizes, with the outputs concatenated along the depth dimension. Factorizations and improvements led to the latest Inception version, including splitting 5×5 convolutions into smaller kernels to reduce computational costs and adding an auxiliary network for better regularization and gradient propagation. Inception-ResNet, the newest variant, replaces standard Inception blocks with residual blocks, introduced by He et al. [10].

2.2 Pair of B-Scans Classification

The goal of the first task of the challenge is to classify the relative evolution of AMD between two consecutive B-scans.

Siamese networks are built by duplicating a CNN backbone to form two identical branches with shared weights. For each pair of B-scans, the B-scan at time t_i is fed into one branch, while the B-scan at time t_{i+1} is input into the other branch. The feature vectors generated by the two branches are concatenated. This concatenated vector is passed through a fully-connected module comprising three linear layers. Each linear layer is followed by ReLU activation, Batch Normalization, and Dropout. This module reduces the dimensionality of the vector to 32. Afterward, the categorical variables are concatenated with this feature vector, which is then passed through two additional linear layers with SiLU activation, producing a 4-dimensional output representing the class logits. To address imbalance, a weighted cross-entropy loss is used, with class weights inversely proportional to the class frequencies. The training process is independently carried out for three different CNN architectures: ConvNext, Inception-ResNetv2, and EfficientNetv2.

In the second phase, an Ensemble of the three trained Siamese networks is created. Each of the three networks, along with their respective categorical fusion modules, is instantiated, loaded with previously trained weights, and frozen to prevent further updates. The final classification layer of each network is removed, and a MobileNetv3 CNN backbone [12] is instantiated to process the 2D infrared localizer image at time t_i, generating a feature vector that is reduced to 16 dimensions by a small fully-connected module. The 32-dimensional feature vectors from the three Siamese networks are concatenated with the 16-dimensional localizer feature vector. The resulting 112-dimensional vector is passed through a fully-connected module composed of 4 linear layers that produces the final classification output.

2.3 Predicting Future AMD Evolution from One B-Scan

For the second task of the challenge, we train three new CNN submodels using Knowledge Distillation, a technique originally designed to transfer knowledge from larger, more complex models to smaller and more efficient models [29]. For this challenge, we adapt the method to predict AMD progression using only the B-scan from time point t_i, aiming to achieve performance comparable to our models that use both t_i and t_{i+1}. The core idea is to leverage the previously trained Siamese CNNs as teacher networks to train new non-Siamese standalone CNNs, enabling them to replicate the learned representations using only the B-scan from t_i. Given the difficulty of the task, the fully connected modules used in Task 1 are expanded, using 7 linear layers before the classification head in each of the individual CNNs. Once trained, the three new CNN models are ensembled using the same approach as previously described.

The Knowledge Distillation framework introduces two additional loss functions alongside the standard cross-entropy loss. The first is the Kullback-Leibler (KL) divergence loss, which helps the student model learn from the teacher class probabilities after applying the softmax function [24]. Since teacher models often produce highly confident probability distributions, the student model might find it difficult to capture the full distribution. To address this, we use a high temperature in the softmax function to soften the probability distribution, making it easier for the student model to learn from it [11]. The KL divergence loss, which measures the difference between the softened teacher distribution P(x) and the student's predicted distribution Q(x), is defined as:

$$\mathrm{KL} = \sum_{x} P(x) \log\left(\frac{P(x)}{Q(x)}\right) \tag{1}$$

The second additional loss function we apply is a cosine similarity loss, which operates on the hidden representations of the student and teacher models [2]. This loss encourages the student model to align its internal representations with those of the teacher model. Since the teacher model has access to both t_i and

t_{i+1}, it has learned richer internal representations than the student model. By minimizing the cosine similarity loss, we encourage the student to mimic the teacher's internal feature space. We choose the layer at the end of the CNNs encoder and before the categorical fusion module as our hidden state vectors.

In addition to the KL divergence and cosine similarity losses, the standard cross-entropy loss is still applied to ensure that the student model learns effectively from the real ground truth labels. The total training loss is a weighted combination of these three losses. Based on our experimental results, we allocate 50% of the loss to cross-entropy (weighted by 0.5), while the remaining 50% is split between the distillation losses, with KL divergence receiving a weight of 0.3 and cosine similarity loss 0.2.

3 Experimental Settings

Data. OCT images were captured using the Spectralis device from Heidelberg Engineering, and sourced from the Ophthalmology department of Brest University Hospital (France). Each OCT stack includes multiple 2D B-scans and a 2D infrared localizer image. B-scans have resolutions of 1536×496, 1024×496, 512×496, or 768×496, while localizer images are 1536×1536 or 768×768. The dataset covers multiple patients, with examinations from different visits over time. Each exam includes an OCT image for each eye, along with categorical data such as patient age and visit number.

For Task 1, the provided ground truth consists of pairwise comparisons between B-scans from the same patient across visits, indicating relative progression as the classes: reduced (0), stable (1), worsened (2), or other (3). There are 68 patients in the dataset, for a total of 14,496 pairwise comparisons. An additional external evaluation set is provided, which includes 34 patients, totaling 7,010 comparisons. For Task 2, a subset of the same patient cohort is used, but only the first B-scan is provided, and the third ("other") class is excluded. There are 61 patients, for a total of 8,082 comparisons. External evaluation for Task 2 includes 29 patients and 3,822 comparisons. We split the training dataset into training, validation, and test sets, at the patient level to prevent data leakage. For Task 1, splits include 50 patients (10,865 comparisons) for training, 9 for validation (1,844 comparisons), and 9 for testing (1,787 comparisons). For Task 2, splits involve 42 patients (5,337 comparisons) for training, 9 for validation (1,461 comparisons), and 10 for testing (1,284 comparisons).

Implementation Details. All experiments use the AdamW optimizer with a weight decay of 0.05 and a batch size of 32, distributed across one or two GPUs on a SR650 node (MCIA platform). The hardware includes two Intel Xeon Gold SKL-6130 processors (6 cores, 2.1 GHz), 192 GB RAM, and 2 NVIDIA P100 GPUs. The training pipeline is built with PyTorch. Images undergo real-time augmentation, including gaussian blur, flips, normalization and random cropping. Image dimensions are resized differently based on the CNN backbone: 384×384 for ConvNext, 256×256 for Inception-ResNetv2, and 224×224 for

EfficientNetv2. Weighted cross-entropy loss is used, with class weights inversely proportional to class frequencies. For Task 1, the weights are 0.0006, 0.0001, 0.0008, and 0.0009 for the classes reduced (0), stable (1), worsened (2), and other (3), respectively. For Task 2, for the three classes in the same order, weights are 0.001, 0.0002, and 0.002.

4 Results and Discussion

All the results reported in this section were obtained using the previously mentioned test sets. The external evaluation was conducted on the separate dataset provided without labels, with the metrics calculation performed online via the challenge platform. We used the same metrics selection, naming conventions, and implementation as provided by the challenge organizers. It is important to note that the F1-score is micro-averaged, and the Rk-correlation metric corresponds to the Matthews correlation coefficient.

Table 1. Performance metrics of our methods for Tasks 1 and 2, evaluated on both the test set and external evaluation set. Metrics reported include F1-score, Specificity, Rk-correlation, and Kappa. The best results for each metric are highlighted in bold.

(a) Task 1

Method	Metrics			
	F1-score	Specificity	Rk-correlation	Kappa
ConvNext(tiny)	0.842	0.923	0.670	0.669
Inception-Resnet(v2)	0.833	**0.926**	0.667	0.663
EfficientNetv2(small)	0.813	0.898	0.594	0.593
Our method	0.860	0.914	0.687	0.685
Our method w/ localizer	**0.862**	0.920	**0.697**	**0.696**
external evaluation Our method w/ localizer	0.832	0.905	0.648	N/A

(b) Task 2

Method	Metrics			
	F1-score	Specificity	Rk-correlation	Kappa
ConvNext(tiny)	0.595	0.702	0.129	0.122
Inception-Resnet(v2)	0.618	0.701	0.104	0.102
EfficientNetv2(small)	0.424	0.682	0.042	0.034
Our method w/ localizer	0.690	**0.722**	**0.140**	**0.138**
external evaluation Our method w/ localizer	**0.693**	0.662	−0.014	0.026

In the absence of prior work on this dataset, we evaluated the results of our Ensemble model directly against the individual baseline CNNs for both tasks. Table 1a shows that the Ensemble method consistently outperforms the baseline models across most metrics, with the exception of Specificity. The Inception-ResNet model exhibits higher specificity compared to other networks. Upon further analysis, this model appears to better distinguish lower-frequency classes.

However, it comes at the expense of the majority group (class 1, stable), which is selected less frequently, leading to a reduced true positive rate. Given the significant influence of class 1 on the overall F1-score, this trade-off ultimately makes the model less competitive. ConvNeXt, which is the most recent and competitive CNN of our method in terms of natural object classification benchmarks, works well for the task 1. Interestingly, this performance advantage does not carry over to Task 2 (Table 1b).

Apart from the B-scans, we also incorporate categorical data and the localizer image. The categorical data is lightweight and does not introduce significant computational overhead, unlike the localizer, for which we used a dedicated encoder. Therefore, to assess its relevance in our approach, we conducted an ablation study. The results (Table 1a) indicate that the localizer slightly improves performance by helping the fusion module select the optimal combination of submodel outputs. However the primary contribution still comes from the Siamese networks, and while the localizer proves beneficial, its use may be questioned given the additional parameters required.

The results from Task 2 reveal a general decline in performance across all models. While the Ensemble method continued to outperform the individual models, the overall performance remains suboptimal. Despite an acceptable F1-score, both the Matthews correlation coefficient and Kappa score indicate poor predictive power. Similar to Task 1, external evaluation suggests a lack of generalization.

Several factors likely contributed to these performance disparities. Task 2 faces a severe class imbalance, with over 80% of the data concentrated in a single class (class 1 - stable). We employed weighted cross-entropy for both tasks, however other methods could have been considered, such as over-sampling or under-sampling, or using specialized loss functions like focal loss, that have demonstrated strong performance on imbalanced distributions [16,21]. Additionally, the sample size is limited, for instance, only 10 unique patients were represented in our test set, and pairs generated from the same patient may introduce correlations, increasing the risk of overfitting to patient-specific features.

Our approach, using Knowledge Distillation (KD), aligns with generalized distillation [18], which extends KD by incorporating the concept of privileged information [13,27]. In this method, the teacher is trained on additional data, and then enables the student to benefit from insights that would otherwise be inaccessible. In our case, our teacher model leverages input B-scans from both t_i and t_{i+1} during training. However, as noted in previous works [5,30], KD typically separates teacher and student training, limiting the ability to share domain knowledge. The teacher may encode information irrelevant to the student, potentially wasting the student's already limited capacity. This one-sided process can lead to inefficiencies, overfitting, or poor generalization, as it assumes that the representations learned by the teacher can be effectively translated to the student's context. Without access to the teacher's additional input data, the student might only mimic the teacher's representations without comprehending them. Motivated by [13], we used representation matching in deeper layers, where high-level features might converge. We also use a weighted combination of three loss components to prevent the student from simply replicating the teacher, and allow it to learn on its own, and therefore adopt teacher representations only when beneficial. This approach could be further improved by leveraging adversarial training or modality-alignment techniques, which have been proposed previously with success in the literature [28,30,32].

In general, future work could explore alternative approaches to pairwise comparisons, such as reconstructing full original examinations for each patient and employing deep ranking methods. Another promising direction would be to implement bagging techniques to address the issue of intra-patient correlation, which could improve generalization.

5 Conclusion

In this paper, we present our deep learning-based solution, a multi-modal Ensemble of Siamese networks, designed to address both tasks of the MARIO challenge. Our model integrates three state-of-the-art CNNs and incorporates categorical patient data along with a 2D infrared OCT image using a custom fusion module. For the second task, we use Knowledge Distillation to train the network to predict disease progression from a single B-scan. Our approach was evaluated on both an internal test set and an external evaluation set. For the first task, we achieved an F1-score of 0.862 and a Matthews correlation coefficient of 0.697 on the internal test set, and an F1-score of 0.832 and an Matthews correlation coefficient of 0.648 on the external evaluation set. The second task proved much more challenging, as our method demonstrated limited predictive power on both sets. This challenge laid the foundation for developing an effective short-term prediction model for neovascular AMD evolution.

Disclosure of Interests. There are no competing interests to declare.

Code Availability. Our code is available on our GitHub: https://github.com/serichard1/neovAMD_siamese_ensemble_MARIO.

References

1. Abdulraheem, M., Oladipo, I.D., Ajagbe, S.A., et al.: Continuous eye disease severity evaluation system using siamese neural networks. ParadigmPlus **4**(1), 1–17 (2023). http://journals.itiud.org/index.php/paradigmplus/article/view/42
2. Aguilar, G., Ling, Y., Zhang, Y., et al.: Knowledge distillation from internal representations. In: Proceedings of the AAAI Conference on Artificial Intelligence, vol. 34, no. 05, pp. 7350–7357 (2020). https://ojs.aaai.org/index.php/AAAI/article/view/6229
3. Anwar, S.M., Majid, M., Qayyum, A., et al.: Medical image analysis using convolutional neural networks: a review. J. Med. Syst. **42**(11), 226 (2018). http://link.springer.com/10.1007/s10916-018-1088-1

4. Arepalli, S., Kaiser, P.K.: Pipeline therapies for neovascular age related macular degeneration. Int. J. Retina Vitreous **7**, 55 (2021). https://www.ncbi.nlm.nih.gov/pmc/articles/PMC8485527/
5. Azad, R., Khosravi, N., Dehghanmanshadi, M., Cohen-Adad, J., Merhof, D.: Medical image segmentation on MRI images with missing modalities: a review (2022). http://arxiv.org/abs/2203.06217. arXiv:2203.06217
6. Bromley, J., Guyon, I., LeCun, Y., et al.: Signature verification using a "Siamese" time delay neural network. In: NIPS'93 (1994)
7. Cai, L., Gao, J., Zhao, D.: A review of the application of deep learning in medical image classification and segmentation. Ann. Transl. Med. **8**(11), 713 (2020). https://www.ncbi.nlm.nih.gov/pmc/articles/PMC7327346/
8. Deng, Y., Qiao, L., Du, M., et al.: Age-related macular degeneration: epidemiology, genetics, pathophysiology, diagnosis, and targeted therapy. Genes Dis. **9**(1), 62–79 (2021). https://www.ncbi.nlm.nih.gov/pmc/articles/PMC8720701/
9. Ganaie, M.A., Hu, M., Malik, A.K., et al.: Ensemble deep learning: a review. Eng. Appl. Artif. Intell. **115**, 105151 (2022). http://arxiv.org/abs/2104.02395. arXiv:2104.02395
10. He, K., Zhang, X., Ren, S., et al.: Deep residual learning for image recognition (2015). http://arxiv.org/abs/1512.03385
11. Hinton, G., Vinyals, O., Dean, J.: Distilling the knowledge in a neural network (2015). http://arxiv.org/abs/1503.02531. arXiv:1503.02531
12. Howard, A., Sandler, M., Chu, G., et al.: Searching for MobileNetV3(2019). http://arxiv.org/abs/1905.02244. arXiv:1905.02244
13. Hu, M., et al.: Knowledge distillation from multi-modal to mono-modal segmentation networks (2021). http://arxiv.org/abs/2106.09564. arXiv:2106.09564
14. Li, M.D., Arun, N.T., Gidwani, M., et al.: Automated assessment and tracking of COVID-19 pulmonary disease severity on chest radiographs using convolutional siamese neural networks. Radiol. Artif. Intell. **2**(4), e200079 (2020). http://pubs.rsna.org/doi/10.1148/ryai.2020200079
15. Li, M.D., Chang, K., Bearce, B., et al.: Siamese neural networks for continuous disease severity evaluation and change detection in medical imaging. NPJ Digit. Med. **3**(1), 1–9 (2020). https://www.nature.com/articles/s41746-020-0255-1
16. Lin, T.Y., Goyal, P., Girshick, R., He, K., Dollár, P.: Focal loss for dense object detection (2018). http://arxiv.org/abs/1708.02002. arXiv:1708.02002
17. Liu, Z., Mao, H., Wu, C.Y., et al.: A ConvNet for the 2020s (2022). http://arxiv.org/abs/2201.03545. arXiv:2201.03545
18. Lopez-Paz, D., Bottou, L., Schölkopf, B., Vapnik, V.: Unifying distillation and privileged information (2016). http://arxiv.org/abs/1511.03643
19. Lopez Pinaya, W.H., Vieira, S., Garcia-Dias, R., et al.: Convolutional neural networks. In: Mechelli, A., Vieira, S. (eds.) Machine Learning, pp. 173–191. Academic Press (2020)
20. Mitchell, P., Liew, G., Gopinath, B., et al.: Age-related macular degeneration. The Lancet **392**(10153), 1147–1159 (2018)
21. Pasupa, K., Vatathanavaro, S., Tungjitnob, S.: Convolutional neural networks based focal loss for class imbalance problem: a case study of canine red blood cells morphology classification. J. Ambient. Intell. Humaniz. Comput. **14**(11), 15259–15275 (2023). https://doi.org/10.1007/s12652-020-01773-x
22. Rasmussen, A., Sander, B.: Long-term longitudinal study of patients treated with ranibizumab for neovascular age-related macular degeneration. Curr. Opin. Ophthalmol. **25**(3), 158–163 (2014). http://journals.lww.com/00055735-201405000-00003

23. Rossi, A., Hosseinzadeh, M., Bianchini, M., et al.: Multi-modal siamese network for diagnostically similar lesion retrieval in prostate MRI. IEEE Trans. Med. Imaging **40**(3), 986–995 (2021). https://ieeexplore.ieee.org/document/9288754/
24. Sun, S., Ren, W., Li, J., Wang, R., Cao, X.: Logit standardization in knowledge distillation. In: Proceedings of the IEEE/CVF Conference on Computer Vision and Pattern Recognition, pp. 15731–15740 (2024)
25. Szegedy, C., Ioffe, S., Vanhoucke, V., et al.: Inception-v4, Inception-ResNet and the impact of residual connections on learning (2016). http://arxiv.org/abs/1602.07261. arXiv:1602.07261
26. Tan, M., Le, Q.V.: EfficientNet: rethinking model scaling for convolutional neural networks (2020). http://arxiv.org/abs/1905.11946. arXiv:1905.11946
27. Vapnik, V., Izmailov, R.: Learning using privileged information: similarity control and knowledge transfer. J. Mach. Learn. Res. (2015)
28. Wang, H., Chen, Y., Ma, C., Avery, J., Hull, L., Carneiro, G.: Multi-modal learning with missing modality via shared-specific feature modelling. In: 2023 IEEE/CVF Conference on Computer Vision and Pattern Recognition (CVPR), Vancouver, BC, Canada, pp. 15878–15887. IEEE (2023). https://doi.org/10.1109/CVPR52729.2023.01524. https://ieeexplore.ieee.org/document/10204754/
29. Wang, L., Yoon, K.J.: Knowledge distillation and student-teacher learning for visual intelligence: a review and new outlooks. IEEE Trans. Pattern Anal. Mach. Intell. **44**(6), 3048–3068 (2022). https://ieeexplore.ieee.org/document/9340578/
30. Wang, Y., et al.: ACN: adversarial co-training network for brain tumor segmentation with missing modalities (2021). http://arxiv.org/abs/2106.14591. arXiv:2106.14591
31. Wong, W.L., Su, X., Li, X., et al.: Global prevalence of age-related macular degeneration and disease burden projection for 2020 and 2040: a systematic review and meta-analysis. Lancet Glob. Health **2**(2), e106–e116 (2014)
32. Zhang, Y., et al.: mmFormer: multimodal medical transformer for incomplete multimodal learning of brain tumor segmentation. In: Wang, L., Dou, Q., Fletcher, P.T., Speidel, S., Li, S. (eds.) MICCAI 2022. LNCS, vol. 13435, pp. 107–117. Springer, Cham (2022). https://doi.org/10.1007/978-3-031-16443-9_11

Author Index

A
Ahsan, Ahmad O. 182
Araújo, Teresa 106
Aresta, Guilherme 106
Arvanitis, Theodoros 64

B
Bekkers, Amerens A. 118
Beltrán-Carrero, Alberto J. 201
Beurton-Aimar, Marie 211
Binder, Patrick 154, 163
Bogunović, Hrvoje 106
Borderie, Laurent 20, 97
Bounias, Dimitrios 192
Budimir, Lovre Antonio 172

C
Caballero-Sastre, Álvaro 201
Cochener, Béatrice 20, 97

D
Ding, Yi 126, 145

E
Elkhouzari, Ayoub 41
Emre, Taha 106
Ertl, Alexandra 192

F
Fischer, Maximilian 192
Forkert, Nils D. 182
Frohmann, Markus 154, 163

G
Galdran, Adrian 54
Gómez-Valverde, Juan J. 201

H
Hanstein, Marlin 192
Hao, Yiding 77

Hauptmann, Florian Max 192
Hernández-Lastras, María J. Aparicio 201
Huber, Marcel 154, 163

J
Jiang, Weili 20, 97

K
Kächele, Jessica 192
Khouani, Amin 30
Kuijf, Hugo J. 118

L
Lachinov, Dmitrii 106
Lamard, Mathieu 20, 97
Le Guilcher, Alexandre 20, 97
Ledesma-Carbayo, María J. 201
Lehmiani, Yasine 41
Li, Yihao 20, 97
Lin, Hui 20, 97
Lončarić, Sven 172

M
Mahmoudi, Abdelhak 41
Maier-Hein, Klaus H. 192
Matovinović, Ivana 172
Matta, Sarah 20, 97
Mazher, Moona 135
Mekki, Ihsane 30
Miao, Wencheng 3

N
Neher, Peter 192
Niederer, Steven A. 135
Nielsen, Christopher 182
Nolden, Marco 192

O
Oghbaie, Marzieh 106
Ou, Chubin 20, 97

P
Pan, Jiangtian 20, 97
Peretzke, Robin 192

Q
Qayyum, Abdul 135
Qin, Pengyao 64
Quellec, Gwenolé 20, 97

R
Razzak, Imran 135
Richard, Sebastien 211

S
Santos-Vicente, Esther 201

T
Tan, Yubo 20, 97
Thirunavukarasu, Arun 64
Torresano-Rodríguez, Javier 201

V
van Liebergen, Nina M. 118
Vršnak, Donik 172

W
Wang, Haoshen 20, 97
Wilms, Matthias 182

X
Xu, Hui 20, 97

Y
Yamagishi, Yosuke 87
Yan, Yangyang 3
Yang, Bo 3

Z
Zhang, Jing 20, 97
Zhang, Le 64
Zhang, Philippe 20, 97

The manufacturer's authorised representative in the EU is Springer Nature Customer Service Centre GmbH, Europaplatz 3, 69115 Heidelberg, Germany. If you have any concerns regarding our products, please contact ProductSafety@springernature.com

Printed and bound by CPI Group (UK) Ltd, Croydon, CR0 4YY

26/03/2026

02078973-0001